Chemistry Research and Applications

nova
science publishers
www.novapublishers.com

Chemistry Research and Applications

Properties and Uses of Antimony
David J. Jenkins (Editor)
2022. ISBN: 979-8-88697-081-4 (Softcover)
2022. ISBN: 979-8-88697-088-3 (eBook)

The Science of Carbamates
Güllü Kaymak (Editor)
2022. ISBN: 978-1-68507-708-2 (Softcover)
2022. ISBN: 978-1-68507-872-0 (eBook)

Deep Eutectic Solvents: Properties, Applications and Toxicity
Carlos Eduardo de Araújo Padilha, PhD, Everaldo Silvino dos Santos, PhD, Francisco Canindé de Sousa Júnior, PhD, Nathália Saraiva Rios, PhD (Editors)
2022. ISBN: 978-1-68507-719-8 (Hardcover)
2022. ISBN: 978-1-68507-799-0 (eBook)

Polycyclic Aromatic Hydrocarbons: Sources, Exposure and Health Effects
Warren L. Gregoire (Editor)
2022. ISBN: 978-1-68507-626-9 (Softcover)
2022. ISBN: 978-1-68507-685-6 (eBook)

Cyanide: Occurrence, Applications and Toxicity
Bill M. Torres (Editor)
2022. ISBN: 978-1-68507-619-1 (Softcover)
2022. ISBN: 978-1-68507-670-2 (eBook)

A Closer Look at Carvacrol
Zak A. Cunningham (Editor)
2022. ISBN: 978-1-68507-627-6 (Softcover)
2022. ISBN: 978-1-68507-634-4 (eBook)

More information about this series can be found at
https://novapublishers.com/product-category/series/chemistry-research-and-applications/

Deepak Kumar Semwal
Editor

The Essential Guide to Alkaloids

nova
science publishers
www.novapublishers.com

https://doi.org/10.52305/KXUM3530

Library of Congress Cataloging-in-Publication Data

ISBN: 979-8-88697-456-0

Published by Nova Science Publishers, Inc. † New York

Contents

Preface

Alkaloid is a class of organic compounds having diverse skeletal structures and exceptional pharmacological properties. These are naturally produced by plants as secondary metabolites to protect themselves from herbivores, pathogens or other environmental stresses. In addition, these are also found in fungi (Psilocybe), animals (toads), insects (ants) and various marine organisms. These are usually basic in nature and have at least one nitrogen atom in their composition. Alkaloids have a long history of therapeutic importance. Many plant families like Menispermaceae, Papaveraceae and Solanaceae are well known for their alkaloidal contents including bisbenzylisoquinoline, morphine and aporphine type alkaloids. Various alkaloids such as morphine, codeine, atropine, caffeine and berberine have a long history of medicinal uses. These are among the leading compounds for novel drug discovery including anticancer (vinblastine, vincristine), antimalarial (quinine), antiasthma (ephedrine), vasodilatory (vincamine), analgesic (morphine) and antidiabetic (piperine) agents. Alkaloid chemistry is one of the major research areas for the past many decades mainly due to the diverse therapeutical effects of alkaloids. In viewing their market demand, many of the alkaloids and their derivatives nowadays are synthesised on a commercial scale. Interestingly, many alkaloids are already used as drugs and some others are under trial.

The present book entitled *The Essential Guide to Alkaloids* is based on general information about alkaloids, their applications mainly in medicine and the chemistry of some natural and synthetic alkaloids. The contents are divided into a total of nine chapters contributed by different subject experts.

The first chapter describes introductory knowledge and pharmaceutical perspectives of alkaloids. This section reported the different classification methods of alkaloids for their better understanding. In addition, various extraction methods and pharmaceutical applications of alkaloids are also included in this part. The second chapter is based on chemical aspects and therapeutic uses of tropane alkaloids. This section covered structural variations, pharmaceutical uses and biosynthesis of tropane alkaloids.

Different uses of alkaloids in traditional medicine are described in the third chapter which is based on scientific and historical evidence. It covers plant-derived alkaloids and their biological activities like acetylcholinesterase inhibitory, anti-inflammatory and antioxidant. Moreover, it covered the activities and mechanisms of various plant-derived alkaloids like tetrahydropalmatine, aloperine, tetrandrine, sinomenine, oxymatrine, galantamine, berberine and harmine. The fourth chapter comprised alkaloids of the genus Solanum such as solamargine, solasonine, tomatidine and kukoamines. The particular sources of these alkaloids include Solanum crinitum, Solanum lycocarpum, Solanum oocarpum and Solanum sisymbriifloium together with their pharmacological activities are also described in this section. Catharanthus roseus, a well-known alkaloid-rich herb is covered in the fifth chapter with its key alkaloids like vinblastine and vincristine. Different pharmacological effects including antioxidant, antimicrobial, antifeedant, anti-sterility, anthelmintic, antidiabetic and antidiarrheal of Catharanthus roseus are described in this section.

The sixth chapter comprised a bromopyrrole alkaloid namely stevensine which is naturally found in marine sponges. Its several pharmacological properties like protein kinase inhibitor, neurological, anti-microbial, anti-tubercular, anti-cancer and anti-parasite are described in this chapter. The seventh chapter focused on alkaloids derived from the Tinospora cordifolia, their structure and function and biosynthesis. This chapter also encompassed the importance of herbal medicine, a general introduction of alkaloids, the biological role of alkaloids in disease targeting and large-scale production of alkaloids using in vitro cultures and elicitor-treatment.

Theophylline, a dimethylxanthine alkaloid with therapeutical properties is covered in the eighth chapter. This chapter also described the applications of theophylline as a bronchodilator, immunomodulator and antiviral which make it an interesting candidate against coronaviruses. In addition to their medicinal properties, alkaloids are also useful as pesticides to protect plants from different pathogens and pests. These applications are included in the ninth chapter with the main emphasis on their unique properties such as being safer for the environment, easily degradable and low toxicity as compared to synthetic pesticides.

The present piece of work is a compilation of both natural and synthetic alkaloids with their diverse biological activities. This piece of work will be useful for postgraduate students for their basic understanding of alkaloids.

Since some of the chapters are based on recent developments in the field of alkaloids, hence this would be helpful for researchers working in the related areas.

Acknowledgments

The quality of the content cannot be verified and maintained without evaluation by the subject matter expert. The following reviewers are acknowledged for their efforts to review the contents of the present book, *The Essential Guide to Alkaloids*.

1. Prof. Vineet Kumar, Forest Research Institute, India
2. Dr. Sekelwa Cosa, University of Pretoria, South Africa
3. Dr. Aijaz Ahmad, University of Witwatersrand, South Africa
4. Dr. Deepak M. Kasote, North Carolina State University, USA
5. Dr. Dinesh Kumar, CSIR-Institute of Himalayan Bioresource Technology, India
6. Dr. Ruchi B. Semwal, Government Postgraduate College Dakpathar, India
7. Dr. Ashutosh Chauhan, Uttarakhand Ayurved University, India
8. Dr. Kumud Upadhyaya, Kumaun University, India

Abbreviations

AChE	acetylcholinesterase
AIDS	acquired immunodeficiency syndrome
Akt	Protein kinase B
API	active pharmaceutical ingredient
C4H	cinnamate-4-hydroxylase
cAMP	cyclic adenosine monophosphate
CDK	cyclin-dependent kinase
CHI	chalcone isomerase
CHIKV	chikungunya virus
CHS	chalcone synthase
CNS	central nervous system
COPD	chronic obstructive pulmonary disease
COVID	coronavirus disease
COX	cyclooxygenase
EEG	electroencephalogram
GBIF	Global Biodiversity Information Facility
GPCRs	G-protein-coupled receptors
GSH	reduced glutathione
HIV	human immunodeficiency virus
IL	interleukin
iNOS	inducible nitric oxide synthase
KLF2	Kruppel-like factor 2
LPS	lipopolysaccharide
MAO	monoamine oxidase
MAO-B	monoamine oxidase-B
MAPK	mitogen-activated protein kinase
MCF-7	Michigan Cancer Foundation-7
MCP-1	monocyte chemoattractant protein-1
MDR	multidrug resistance
MIP	macrophage inflammatory protein

NDDs	neurodegenerative disorders
NF-kB	nuclear factor kappa
NMDA	N-methyl D-aspartate
NMR	nuclear magnetic resonance
NO	nitric oxide
NSAIDs	nonsteroidal anti-inflammatory medications
PAL	phenylalanine ammonia-lyase
PEG	prostaglandin
RANKL	receptor activator of nuclear factor kappa-B ligand
Rhy	rhynchophylline
RNS	reactive nitrogen species
ROS	reactive oxygen species
SARS-CoV-2	severe acute respiratory syndrome coronavirus-2
SOD	superoxide dismutase
TAL	tyrosine ammonia-lyase
TGF	transforming growth factor
TIAs	terpenoid indole alkaloids
TLR4	Toll-like receptor 4
TMV	tobacco mosaic virus
TNF	tumour necrosis factor
U87-MG	Uppsala-87 malignant glioma
VCAM-1	vascular cell adhesion molecule-1
3CLpro	chymotrypsin-like protease protein
4CL	4-coumarate-CoA-ligase

Chapter 1

Alkaloids: Introductory Knowledge and Pharmaceutical Perspectives

Ravindra Semwal[1,*], Ankit Kumar[1], Sunil Kumar Joshi[2] and Deepak Kumar Semwal[3]

[1]Research and Development Centre, Faculty of Biomedical Sciences, Uttarakhand Ayurved University, Harrawala, Dehradun, India
[2]Uttarakhand Ayurved University, Harrawala, Dehradun, India
[3]Department of Phytochemistry, Faculty of Biomedical Sciences, Uttarakhand Ayurved University, Harrawala, Dehradun, India

Abstract

Since ancient times, people have used plants containing alkaloids as remedies in the form of teas, poultices, potions, etc. to treat some diseases or sometimes as poison and dyes. They are generally basic compounds having a bitter taste and contained nitrogen in their structure. The widespread distribution of alkaloids and their wide array of structures makes their classification often difficult, hence, there are different classification methods are been reported for better understanding. There are various extraction and isolation methods, available to obtain alkaloids in a smooth manner, where some are common and some are specific. Because of the medicinal use of alkaloids, proper identification is needed for safety purposes for which, a number of identification tests are available. In modern days, the biological significance of alkaloids depends on their significant relationship with health benefits. In general, many alkaloids are pharmacologically active and possess various physiological activities in humans and animals. Herein, this work is a comprehensive revision of the web of knowledge on the introductory

* Corresponding Author's Email: ravindra.semwal@gmail.com.

In: The Essential Guide to Alkaloids
Editor: Deepak Kumar Semwal
ISBN: 979-8-88697-456-0

sight of alkaloids, their classification, extraction techniques, identification tests and pharmaceutical applications.

Keywords: quinoline alkaloids, tropane alkaloids, Solanaceae, pharmaceutical application, natural alkaloids

1. Introduction

Alkaloids are traditionally known for their bitter taste, basicity, plant origin, and physiological actions. The term "alkaloid" was first introduced by German scientist, Carl Friedrich Wilhelm Meissner in 1819. The term alkaloid was originally used to refer to base-type compounds that contained nitrogen and react with acids forming salts. The word alkaloid is derived from the Arabic word "al-qali" (Goyal, 2013; Roberts, 2013). The use of alkaloids containing plants as drugs, dyes or poisons, can be traced back almost to the beginning of civilization. Curare alkaloids, having muscle relaxant properties, are the active ingredients of arrow poisons used by South American Indians since ancient times. Opium derived from *Papaver somniferum* L. was used as medicine in ancient Greece, Egypt and Rome (Jing et al., 2014; Croteau et al., 2000). Chemically, alkaloids are defined as crystalline, colourless substances with a bitter taste that can form salts when being united with acids, however, in plants, they can hide in a free state, like salts or like N-oxides (Kutchan, 1995; O'Connor, 2010; Amirkia and Heinrich 2014; Encyclopedia Britannica 2018; Bribi 2018). They are pharmacologically active compounds which are consists of two or more fused organic compounds including hetero-atoms in it, which determine the properties of alkaloids. The presence of at least one nitrogen atom is a general chemical feature of alkaloids (Roberts and Wink, 1998; Dostal, 2000). These nitrogen atoms are found as primary or another form of amines and may provide basicity to the alkaloid. They may also contain some neutral or weakly acidic compounds (Manske and Holmes, 2014; McNaught and McNaught, 1997). Alkaloids are found mainly in plants, but to a lesser extent, they are also found in animals such as muscopyridine from the musk of a deer, castoramine from North American rodents. They are also found in insects, marine invertebrates, and microorganisms, i.e., pyocyanin from *Pseudomonas aeruginosa* (Lovell et al., 1986; Lu et al., 2012; Roberts, 2013; Bribi, 2018). Some synthetic compounds are also considered alkaloids too and are more effective as compared to naturally occurring alkaloids (Lewis, 1998). Many alkaloids are toxic in nature, but they often have pharmacological potential if they will be used up to the dose specified for a particular purpose. Some alkaloids are been used medicinally since ancient times such as ephedra, opium poppies, and cocoa leaves. The study of alkaloids came into light in 1806, when morphine was isolated from opium by scientist Serturner. The alkaloids were first isolated successfully in the nineteenth century, and alkaloid-containing drugs were marketed (Ladenburg, 1881), however, the structure of the first alkaloid, namely coniine, was recognized in the year 1870 by scientist Schiff.

Apart from carbon, nitrogen, or hydrogen, alkaloids may comprise sulfur and rarely bromine, phosphorus, or chlorine (Knunyants and Zefirov, 1988). Alkaloids are optically active, except papaverine they are bitter in taste, except coniine they are levorotatory, except berberine-yellow, harmaline-reddish and betanidine-reddish they are colourless, except nicotine and coniine alkaloids are crystalline solids having solubility in organic solvents. They are basic in nature and form salts with acids. Some of the alkaloids that form salts are called quaternary amines i.e., cinchona alkaloids with quininic acid, while some exist as free bases i.e., nicotine. Alkaloids also exist as glycosides such as solanum alkaloids and esters like atropine. The biological activity of the alkaloids generally depends on the protonation in which the amine function is transferred into a quaternary amine at physiological pH.

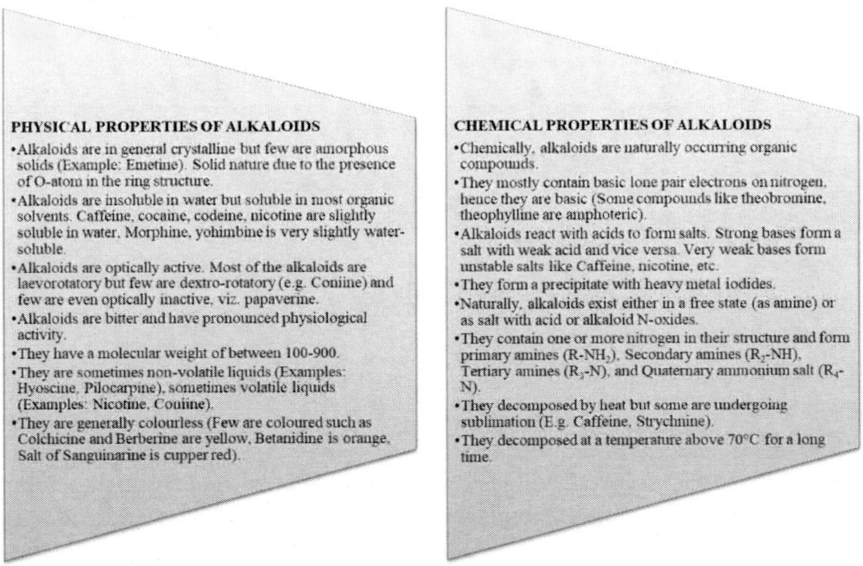

Figure 1. Physical and chemical properties of alkaloids.

Alkaloids are mostly produced by higher plants and they are procured from various parts of these plants such as the whole plant of Datura, the bark of Cinchona, seeds of Nux-vomica, roots of Aconite, leaves of Tobacco, fruits of Black pepper, the latex of Opium, etc. The common physical and chemical properties of alkaloids are given in Figure 1.

Alkaloids can be present in different organs of the cell like mitochondria, vesicles, chloroplasts, and vacuoles and its precursors are derivatives of metabolic pathways, such as glycolysis. Figure 2 shows the biosynthetic pathways of the group of alkaloids which is partially adopted from wink, 2010 (Wink, 2010; O'Connor, 2010; Aniszewski, 2015). Alkaloids are abundant in nature, i.e., they are found in at least 25% of plants. They are generally produced to facilitate the survival of plants in the ecosystem because they are allelopathic compounds and have the potential to be a natural herbicide (Jing et al., 2014; Mazid et al., 2011).

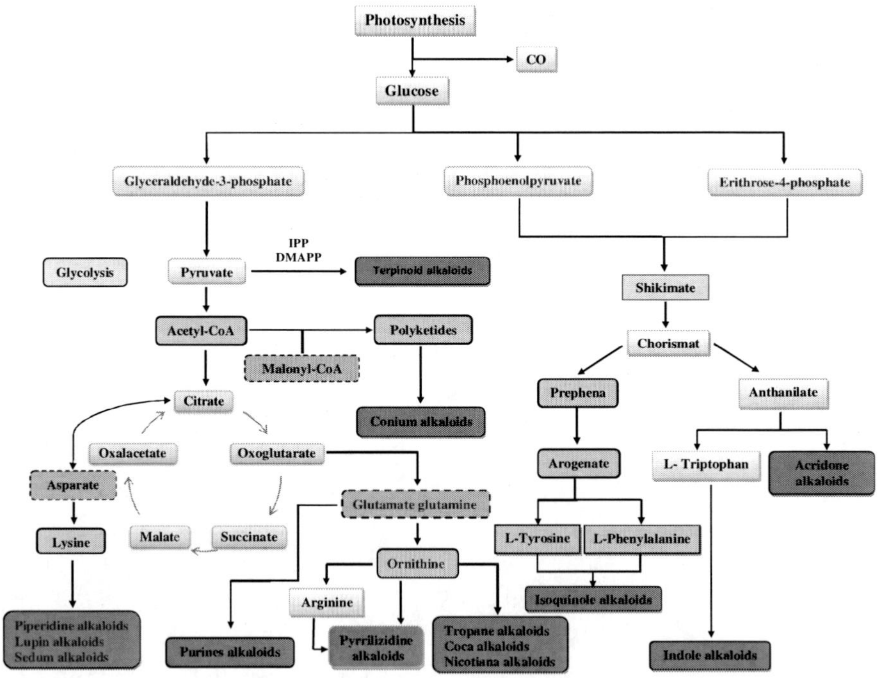

Abbreviations: IPP - isopentenyl diphosphate; DMAPP - dimethylallyl diphosphate. (This figure is partially adopted from Gutierrez-Grijalva et al., 2020).

Figure 2. Metabolic pathway of alkaloid biosynthesis.

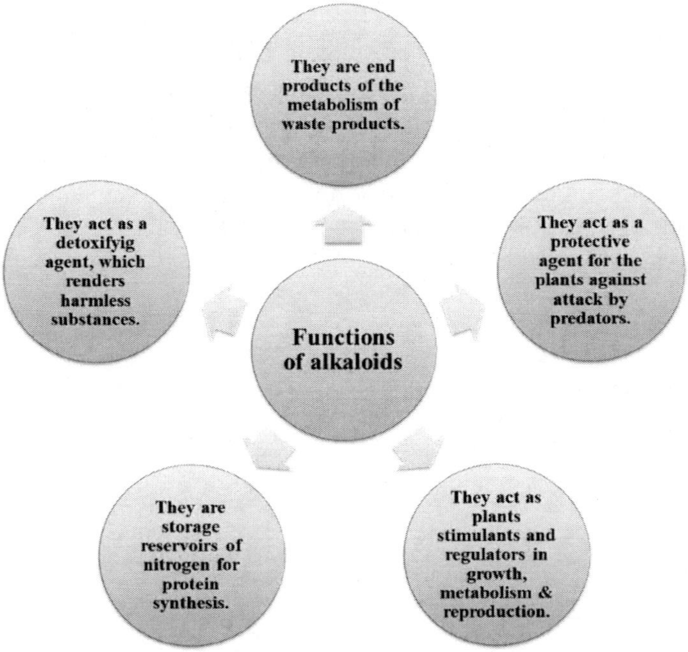

Figure 3. Different functions of alkaloids.

Nowadays, researchers have identified several types of alkaloids in over 4000 different plants. Some plant families are more popular and are known for their high content, such as the Papaveraceae, Ranunculaceae, Solanaceae, Amaryllidaceae, etc. (*Encyclopedia Britannica*, 2018). The common functions of alkaloids are given in Figure 3.

2. Classification of Alkaloids

Classification of the alkaloid depends upon many characteristics such as chemical properties, taxonomy, nature/basicity and pharmacological features. Apart from the position of nitrogen, the classification also depends on their biosynthetic pathway, i.e., terpenoid indole alkaloids, benzylisoquinoline alkaloids, tropane alkaloids, and purine alkaloids (Facchini 2001). Alkaloids are a large and complex group of highly diverse natural products, hence are classified using different techniques. The pharmacological classification is based on the clinical use or pharmacological activity i.e., analgesic and cardioactive alkaloids. Taxonomic classification is based on the family or genus i.e., rauwolfia and cinchona alkaloids. Biosynthetic classification is also based on the type of precursors or building block compounds used by plants to synthesize alkaloids i.e., morphine, papaverine, narcotine and colchicine may be listed as phenylalanine- and tyrosine-derived bases. The last one, chemical classification, is based on the chemical structure of the alkaloid i.e., thebaine, for example, is an isoquinoline derivative alkaloid of opium; aspidospermine is an indole alkaloid (Pelletier, 1983). These groups are shown in Figure 4.

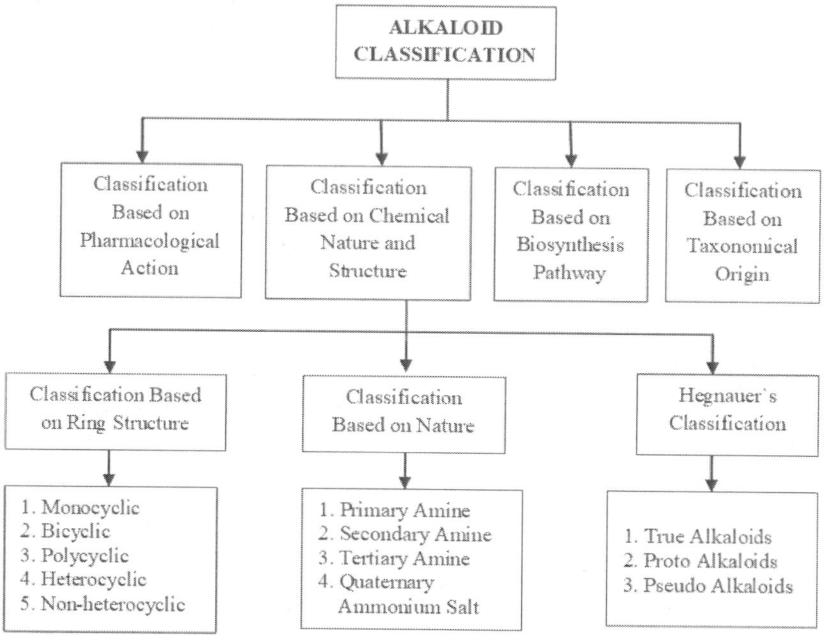

Figure 4. A flow chart showing the major classification of alkaloids.

2.1. Classification Based on Chemical Nature and Structure

This is the popular classification of alkaloids, based on the position of nitrogen in the alkaloid structure and whether it is s part of the ring or not. In this regard, alkaloids are identified as heterocyclic alkaloids or typical alkaloids and non-heterocyclic or atypical alkaloids (Evans, 2009). According to chemical nature and structure, alkaloids are further classified into several types as given in Table 1, 2 and 3.

Table 1. Monocyclic, bicyclic and polycyclic alkaloids

Type of alkaloids	Example	Structure	Attributes
Monocyclic alkaloids	nicotine, etc.	Figure 5	They contain a single, unfused ring.
Bicyclic alkaloids	atropine, cocaine, etc.	Figure 5	They consist of molecules with a 1,4 nitrogen-bridged cycloheptane structure.
Polycyclic alkaloids	strychnine, cannabinol, morphine, codeine, etc.	Figure 5	They are having more than two rings.

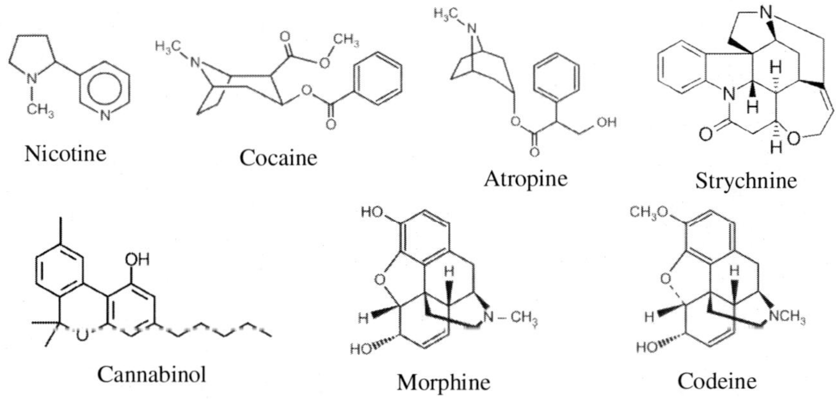

Figure 5. Chemical structures of monocyclic, bicyclic and polycyclic alkaloids.

2.1.1. Heterocyclic Alkaloids

This is the most comprehensively established classification, based on the presence of a basic heterocyclic nucleus in their structure. The heterocyclic alkaloids are divided into pyrrole, pyrrolidine, pyrrolizidine, pyridine, piperidine, tropane, quinolone, isoquinoline, aporphine, quinolizidine, indole, indolizidine, and imidazole. Alkaloids, such as berberine, salsoline, geissospermine, piperine, nicotine, lobeline, nantenine, cocaine, quinine, dopamine, and morphine can be found within these groups (Hussain et al., 2018; Evans, 2009). Types of alkaloids, examples, biological sources, family, and uses of some heterocyclic alkaloids are shown in Table 2 and common skeletons of different alkaloids are given in Figure 6.

Table 2. Different heterocyclic alkaloids and their occurrence

Types of alkaloids	Example	Attributes	References
Tropane alkaloid	cocaine, atropine, scopolamine and their derivatives	Abundantly found in the Solanaceae family. Derived from ornithine and acetoacetate. Pyrrolines are the precursor of these types of alkaloids. Most of them are esters of mono, di, trihydroxytropane, having a wide range of hydroxylation arrangements.	Biastoff and Drager, 2007; Gossauer, 2003; Hemscheid, 2000
Piperidine alkaloids	lobeline	Presence of odor is a common feature of these alkaloids. They exert chronic neurotoxicity. They are lysine-derived alkaloids but some of them are also derived from acetate, acetoacetate, in an analogous fashion to the simple pyrrolidine alkaloids.	Strunz and Findlay, 1985
Quinolines alkaloid	cinchonine, cinchonidine, quinine and quinidine	Achieved exclusively from the bark of the Cinchona plant. But a variety of simple heteroaromatic quinolines are also isolated from various marine sources (4,8-quinolinediol from cephalopod ink and 2-heptyl-4- hydroxyquinoline from a marine pseudomonad).	Michael, 2008; Soares et al., 2007
Isoquinoline alkaloids	morphine, codeine, salsoline, mimosamycin and reticuline	Derived from tyrosine or phenylalanine. Made from a predecessor of dopamine (3,4-dihydroxytryptamine) associated with a ketone or aldehyde.	Baker, 1996; Bentley, 2006; Schiff Jr, 1987
Indole alkaloids	aplysinopsin, gramine, indirubin, 6,60-dibromoindigotin, tryptamine and ergotamine	Derived from tryptophan. Polyhalogenation is a common feature of these alkaloids.	Fresneda and Molina, 2004; Kam and Choo, 2006; Kawasaki and Higuchi, 2005; Knolker, 2008

Types of alkaloids	Example	Attributes	References
Steroidal alkaloids	aminopregnanes, cyclone-samandione, jervine, soladulcidine, rubijervine and bufotoxin	Typically originated from higher plants, which belong to Liliaceae, Solanaceae, Apocynaceae, and Buxaceae families, but some are also isolated from amphibians too.	Ata and Andersh, 2008; Dey et al., 2018; Habermehl, 1967; Keeler, 1986; Ripperger, 1998; Tomko and Voticky, 1973; Zhao et al., 2015
Imidazole alkaloid	pilocarpine	The imidazole ring of these alkaloids is previously made at the stage of the precursor, so they are an exemption in the transformation procedure of structures.	Jin, 2006; Maat and Beyerman, 1984
Purine alkaloids	caffeine, theophylline and theobromine	Purine is the nitrogenous base of nucleotide (building block of DNA and RNA), which consists of purine ring and pentose sugar along with another base pyrimidine.	Ashihara et al., 2013; Lean et al., 2011; Rosemeyer, 2004
Pyrrolidine alkaloids	hygrine and brevicolline	Derived from arginine and lysine with the addition of acetate/malonate units. Found in families such as Fabaceae, Erythroxylaceae, and Moraceae.	Cheeke, 1988; Robertson and Stevens, 2014
Pyrrole alkaloids	derivatives of heme and chlorophyll	Found mostly in the Rutaceae family, within the genera Murraya, Glycosmis, and Micromelum	Estevez et al., 2014; Bauer and Knolker 2012
Pyrrolizidine alkaloids	senecionin, acetylindicin, heliotrine, heliofoline, indicine, indicinine and lasiocarpine	Abundantly found in the Asteraceae and Fabaceae families of plants. The majority of pyrrolizidine alkaloids occur in plants as N-oxides, whose role is lost during the isolation process. They have extensively reviewed alkaloids because of their toxic effects, especially liver damage.	Koleva et al. 2012; Schramm et al., 2019, Robins, 1982; Schardl et al., 2007; Wrobel, 1985

Table 2. (Continued)

Types of alkaloids	Example	Attributes	References
Pyridine alkaloids	piperine, coniine, trigonelline, arecoline, arecaidine, guvacine, cytisine, lobeline, nicotine, anabasine, sparteine and pelletierine	Derivatives of the amino acid L-ornithine. Found in the botanical families Aizoaceae, Annonaceae, Apocynaceae, Araceae, Bignoniaceae, Dipsacaceae, Gramineae, Palmae, and Umbelliferae.	Kaur and Arora 2015; Silva Teles et al., 2019
Aporphine alkaloids	caaverine, lirinidine, asimilobine, n-methyl-asimilobine, nornuciferine, nuciferine, anonaine, magnoflorine, dicentrine, boldine, galucine and neolitsine.	Derived from isoquinoline alkaloids. Isolated from approximately 100 genera and diverse families (approximately 20), such as Annonaceae, Menispermaceae, Papaveraceae, Ranunculaceae, Lauraceae, Monimiaceae, Magnoliaceae, and Berberidaceae, among others.	Chen et al., 2013; Muthna et al., 2013
Quinolizidine alkaloids	lupanine, cytosine and sparteine	Derivatives of the amino acid L-lysine mainly found in the Leguminosae family.	Bunsupa et al., 2012; Szoke et al., 2013; Kaur and Arora 2015
Indolizidine alkaloids	swainsonine, castanospermine and lentiginosine	Derivatives of L-lysine. Found in different plants of the genus Ipomoea.	Meira et al., 2012; Diaz 2015; Michael 2008

Table 3. Different non-heterocyclic alkaloids and their sources

Types of alkaloid	Structure	Example	Biological source & family	Uses
Pheny lethylamine	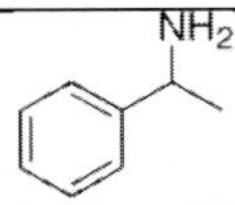	Ephedrine from Ephedra	Ephedra sinica (Ephedraceae)	Asthma and stimulant
Steroidal	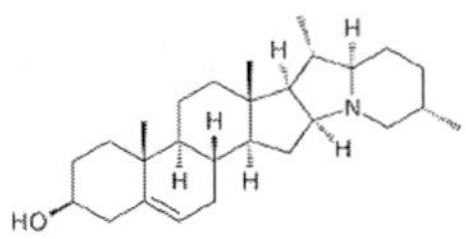	Connesine from kurchi	Holarrhena antidysenterica (Apocynaceae)	Antidysentric
Tropolone	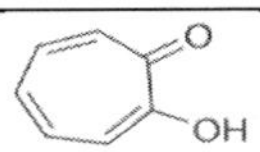	Colchicine from Colchicum	Colchicum autumnale (Colchicaceae)	Gout

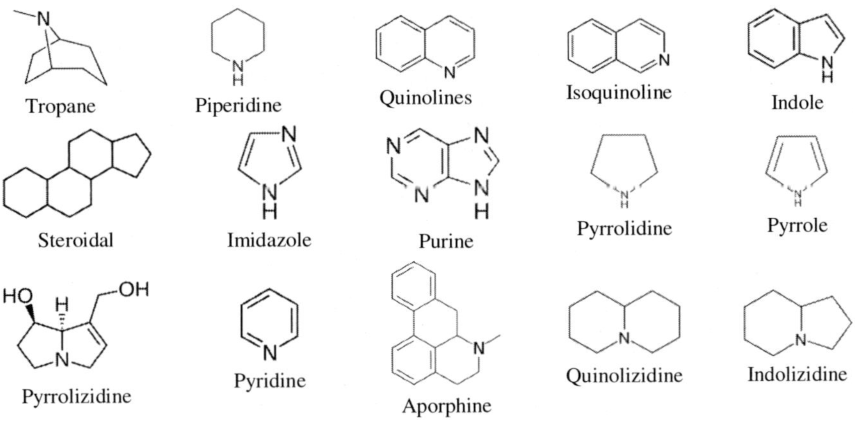

Figure 6. Common skeletons of different alkaloids

2.1.2. Non-Heterocyclic Alkaloids

Types of alkaloids, structure, example, biological sources, family, and uses of non-heterocyclic alkaloids are shown in Table 3 (Bribi et al., 2017; Evans, 2009).

2.2. Classification Based on Nature or Basicity

The basicity of alkaloids is more in R_2-NH alkaloids as compared to R-NH_2, which are more basic than R_3-N. The basicity of alkaloids is due to the presence of a lone pair of electrons on nitrogen. The basicity increases if the adjacent group is electron-releasing like alkali whereas the basicity decreases if the adjacent group is electron-withdrawing like carbonyl and amide group (Cordell, 1991; Evans, 2009). As per nature/basicity, alkaloids are classified as given in Figure 7.

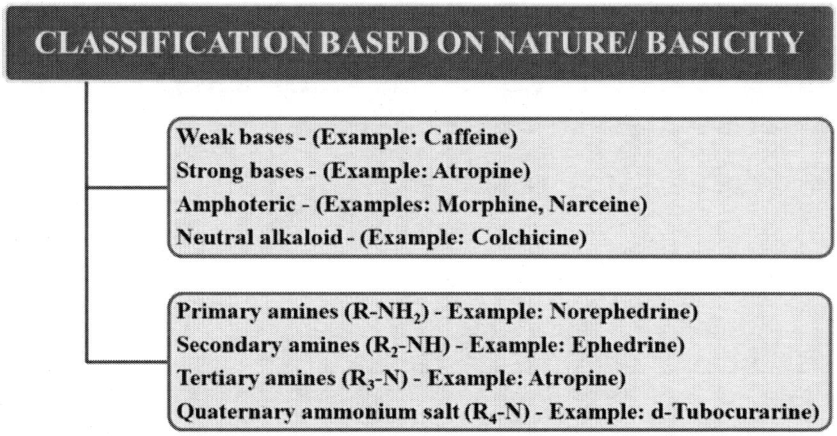

Figure 7. Classification of alkaloids based on nature/basicity.

2.3. Classification Based on Hegnauer's Method

Hegnauer divided alkaloids into three major classes i.e., true alkaloids, protoalkaloids, and pseudoalkaloids (Hegnauer, 1963).

2.3.1. True Alkaloids

True alkaloids are chemically complex, physiologically active compounds, derived directly from cyclic amino acids and having nitrogen in their heterocyclic ring. They are basic in nature due to the presence of a lone pair of electrons on the nitrogen atom and reveal strong toxicity in edible use. True alkaloids are found in plants in the free state, as salts form, and as N-oxides. They can be found in nature forming salts with some organic acids, such as oxalic, lactic, malic, tartaric, acetic, and citric acid (Talapatra, 2015; Henning 2013). The primary precursors of these alkaloids are viz. amino acids (like l-ornithine, l-lysine, l-phenylalanine/l-tyrosine, l-tryptophan, and l-histidine), quinine, morphine, atropine, etc. (Kukula-Koch and Widelski, 2017). These precursors are the basis of a number of alkaloids such as tryptophan is the base of the indole, quinoline, and pyrroloindole alkaloids; tyrosine is the basis of the isoquinoline alkaloids; ornithine is the basis for tropane, pyrrolizidine, and pyrrolidine alkaloids; and lysine for the quinolizidine and piperidine alkaloids; aspartate is the base for the pyridine alkaloids; anthranilic acid is the precursor for quinazoline, quinoline, and acridone alkaloids; and the derivatives of histidine are the imidazole alkaloids (Bottger et al., 2018; Kaur et al., 2019; Aniszewski, 2015).

2.3.2. Proto Alkaloids

Alkaloids do not have nitrogen in their heterocyclic ring and contain the nitrogen atom outside the ring, which remains as part of a side chain, and are considered proto alkaloids. They are mostly derived from amino acids or biogenic amines (such as L-tryptophan, phenylalanine, and L-tyrosine), ephedrine, hordenine, mescaline, etc. Protoalkaloids derived from L-tyrosine and L-tryptophan, are further derived into phenylethylamine and terpenoid indole, respectively (Alves de Almeida, et al., 2017). Mescaline is the most common phenylethylamine alkaloid which can be obtained from Lophophora williamsii, also known as peyote (Beyer et al., 2009). However, monoterpenoid indole alkaloids are a large group of alkaloids in which around 3000 alkaloids have been identified in families such as Apocynaceae, Loganiaceae, and Rubiaceae (Pan et al., 2016). Ephedrine, colchicine, cathinone, etc. are a few more examples of protoalkaloids but they are not so common (Wansi et al., 2013; Talapatra and Talapatra, 2015; Jayakumar and Murugan, 2016; Kukula-Koch and Widelski, 2017).

2.3.3. Pseudo Alkaloids

They are not derived from amino acids but derived from the precursors of amino acids. Generally, they are derivatives of acetate, pyruvic acid, adenine/guanine, or geraniol (Aniszewski 2015). They acquire nitrogen through the transamination process and become a part of the heterocyclic ring, thus nitrogen atom is inserted into the heterocyclic ring at a late stage. They are said pseudo alkaloids as they do not give a positive test for common tests for alkaloids. Examples of these classes include caffeine, coniine, capsaicin and diterpenoid alkaloids obtained from a variety of sources such as the genus Aconitum, Consolida and Delphinium (Wang et al., 2010; Gao et al., 2012). Polyamine (derivatives of putrescine, spermidine and spermine), peptide and cyclopeptide are some other types of alkaloids (Table 4).

Table 4. True, proto, pseudo, polyamine and peptide alkaloids and their examples

Class	Examples
True alkaloids (alkaloids with nitrogen heterocycles)	
Pyrrolidine derivatives	cuscohygrine, hygrine, hygroline and stachydrine
Tropane derivatives	atropine, scopolamine, hyoscyamine, cocaine and ecgonine
Pyrrolizidine derivatives	retronecine, heliotridine, laburnine, indicine, lindelophin, sarracine, platyphylline, trichodesmine, loline, n-formylloline and n-acetylloline
Piperidine derivatives	sedamine, lobeline, anaferine, piperine, coniine and coniceine
Quinolizidine derivatives	lupinine, nupharidin, cytisine, sparteine, lupanine, anahygrine, matrine, oxymatrine, allomatridine, ormosanine and piptantine
Indolizidine derivatives	swainsonine and castanospermine
Pyridine derivatives	trigonelline, ricinine, arecoline, nicotine, nornicotine, anabasine, anatabine, actinidine, gentianine, pediculinine, evonine, hippocrateine and triptonine
Isoquinoline derivatives and related alkaloids	salsoline, lophocerine, n-methylcoridaldine, noroxyhydrastinine, cryptostilin, ancistrocladine, papaverine, laudanosine, sendaverine, cularine, yagonine, argemonine, amurensine, cryptaustoline, berberine, canadine, ophiocarpine, mecambridine, corydaline, hydrastine, narcotine (noscapine), fumaricine, emetine, protoemetine, ipecoside, sanguinarine, oxynitidine, corynoloxine, glaucine, coridine, liriodenine, pronuciferine, glaziovine, kreysiginine, multifloramine, bulbocodine, morphine, codeine, thebaine, sinomenine, kreysiginine, androcymbine, imerubrine, rufescine, imeluteine, lycorine, ambelline, tazettine, galantamine, montanine, erysodine, erythroidine, atherosperminine, protopine, oxomuramine, corycavidine and doriflavin
Oxazole derivatives	annuloline, halfordinol, texaline and texamine
Isoxazole derivatives	ibotenic acid and muscimol
Thiazole derivatives	nostocyclamide and thiostreptone
Quinazoline derivatives	febrifugine, glycorine, arborine, glycosminine and vazicine (peganine)
Acridine derivatives	Rutacridone and acronicine
Quinoline derivatives	cusparine, echinopsine, evocarpine, flindersine,dictamnine, fagarine, skimmianine, quinine, quinidine, cinchonine and cinhonidine

Class	Examples
Indole derivatives	Non-isoprene indole alkaloids- serotonin, psilocybin, dimethyltryptamine, bufotenin, harman, harmine, harmaline, eleagnine, physostigmine (eserine), etheramine, physovenine and eptastigmine Semiterpenoid indole alkaloids- ergotamine, ergobasine, ergosine Monoterpenoid indole alkaloids- ajmalicine, sarpagine, vobasine, ajmaline, yohimbine, reserpine, mitragynine, strychnine, brucine, aquamicine, vomicine, ibogamine, ibogaine, voacangine, vincamine, vinca alkaloids, vincotine and aspidospermine
Imidazole derivatives	histamine, pilocarpine, pilosine and stevensine
Purine derivatives	caffeine, theobromine, theophylline and saxitoxin
Proto alkaloids (alkaloids with nitrogen in the side chain)	
β-Phenylethylamine derivatives	tyramine, ephedrine, pseudoephedrine, mescaline, cathinone, catecholamines (adrenaline, noradrenaline and dopamine)
Colchicine alkaloids	colchicine and colchamine
Muscarine	muscarine, allomuscarine, epimuscarine and epiallomuscarine
Benzylamine	capsaicin, dihydrocapsaicin, nordihydrocapsaicin and vanillylamine
Pseudo alkaloids (terpenes and steroids)	
Diterpenes	aconitine and delphinine
Steroidal alkaloids	solanidine, cyclopamine and batrachotoxin
Polyamines alkaloids	
Putrescine derivatives	paucine
Spermidine derivatives	lunarine and codonocarpine
Spermine derivatives	verbascenine and aphelandrine
Peptide (cyclopeptide) alkaloids	
Peptide alkaloids with a 13-membered cycle	nummularine C, nummularine S, ziziphine A and sativanine H
Peptide alkaloids with a 14-membered cycle	frangulanine, scutianine J, scutianine A, integerrine, discarine D, amphibine F, spinanine A, amphibine B and lotusine C
Peptide alkaloids with a 15-membered cycle	mucronine A

2.4. Classification Based on Pharmacological Action

Based on pharmacological activity (Guggisberg and Hesse, 2001; Bribi, 2018), alkaloids are classified into various groups as shown in Table 5.

Table 5. Classification of alkaloids based on their pharmacological actions

Pharmacological activity	Alkaloid source plant	Biological source	Alkaloids examples
Narcotic analgesic	Opium	*Papaver somniferum* (Papaveraceae)	morphine, codeine
CNS Stimulant	Tea	Thea sinensis (Theaceae)	caffeine
	Nux vomica	*Strychnous nuxvomica* (Loganaceae)	strychnine
Anticancer	Vinca	*Catharanthus roseus* (Apocynaceae)	vincrystine, vinblastine
	Taxol	*Taxus brevifolia* (Taxaceae)	paclitaxel
Antihypertensive	Rauwolfia	*Rouwolfia serpentine* (Apocynaceae)	reserpine
Bronchodilator	Ephedra	*Ephedra gerardiana* (Ephedraceae)	ephedrine
	Vasaka	*Adhatoda vasica* (Acanthaceae)	vasicinone
Smooth muscle relaxant	Belladona	*Atropa belladonna* (Solanaceae)	atropine
Antitussive	Opium	*Papaver somniferum* (Papaveraceae)	codeine
Mydriatics	Belladona	*Atropa belladonna* (Solanaceae)	atropine
Myotics	Pilocarpus	*Pilocarpus jaborandi* (Solanaceae)	pilocarpine
Antiparacitics	Cinchona	*Cinchona calisaya* (Rubiaceae)	quinine
Local anaesthetic	Coca	*Erythroxylum coca* (Erythroxylaceae)	cocaine
Antiarrhythmic	Cinchona	*Cinchona calisaya* (Rubiaceae)	quinidine

2.5. Classification Based on Biosynthesis Pathway

Based on biosynthesis (Eguchi et al., 2019), the alkaloids are classified as given in Table 6.

Table 6. Classification of alkaloids based on their biosynthesis pathways

Precursor	Type of alkaloid	Example
Ornithine derived	pyrrolidine	nicotine
	tropane	atropine, cocaine
Lysine derived	piperidine and pyridine	coniine, lobeline
	quinazolidine	lupinine
Tyrosine derived	isoquinoline	morphine, codeine, berberine
	amino	colchicine
Tryptophan derived	indole	ergot, vincristine, reserpine, strychnine
	quinoline	cinchona, quinine, quinidine
Histidine derived	imidazole	pilocarpine
Phenylalanine derived	amino alkaloid	ephedrine

2.6. Classification Based on Taxonomical Origin

Alkaloids are classified based on biological sources. Examples: Quinine from the bark of *Cinchona calisaya*, Rauwolfia from roots of *Rauwolfia serpentina*, Morphine from dried latex of *Papaver somniferum*, etc. (Bruneton, 1999). The Rubiaceae family includes iridoids, indole alkaloids, anthraquinones, terpenoids (diterpenes and triterpenes), flavonoids and other phenolic derivatives alkaloids (Farias, 2006). Vinca alkaloids are popular as anticancer agents and some of them are available on market for clinical use, i.e., vinblastine, vinorelbine, vincristine and vindesine (Kufe, et al., 2003). Solanaceae is also a widespread family of plants containing tropane alkaloids, glycoalkaloids, pyrrolizidine and indole alkaloids.

Table 7. Classification of alkaloids based on their botanical origin

Botanical origin	Examples
Papaver (opium) alkaloids	morphine, thebaine, papaverine and narcotine
Cinchona alkaloids	cinchonine, cinchonidine, quinine, and quinidine
Rauvolfia alkaloids	ajmalicine, ajmaline, reserpine, rescinnamine, yohimbine, and serpentine
Catharanthus alkaloids	vinblastine, vinorelbine, vincristine and vindesine
Strychnos alkaloids	strychnine and brucine
Ergot alkaloids	dihydroergotamine and ergotamine
Cactus alkaloids	mescaline, hordenine and camegine
Solanum alkaloids	chaconine, solanine, solasodine, tomatidine, tomatine and solanidine

Belladonna, Datura, Henbane, Mandrake, Tobacco, etc. are some examples of alkaloids belonging to the Solanaceae family (Jerzykiewicz, 2007). Similarly; there is a number of alkaloid families which are been used clinically and some are in clinical trials. Based on botanical origin, the alkaloids classification is given in Table. 7.

3. Identification Tests for Alkaloids

There are mainly two types of tests to identify alkaloids, precipitation reaction tests and colour reaction tests. Some of these tests are common for all kinds of alkaloids and some are specific. The popular identification tests are briefly described in Table 8.

Table 8. Different identification tests for alkaloids

Test name	Method
Colour reactions	
Vitali Morin test	The alkaloid sample is mixed with fuming nitric acid and evaporated to dryness. Dissolve the residue in acetone and add a methanolic solution of KOH. It gives a violet colour. It indicates the presence of the tropane group. The test is mainly performed for Solanaceae family drugs such as Belladonna, Datura, Henbane, Mandrake and Tobacco.
Van Urk's test	The alkaloid sample reacts with para-dimethyl amino benzaldehyde and dilute sulphuric acid and gives blue colour. It is applicable to find the presence of the indole group especially the clavinet group of alkaloids. Ergot alkaloids can be identified by this test.
Friend's test	The sample reacts with Froehd's reagent (sodium molybdate in concentrated sulphuric acid) and forms brownish-black colour due to the presence of the isoquinoline group. Opioid alkaloids can be identified by this test.
Thalloquin test	Cinchona powder reacts with bromine water in presence of strong ammonia and gives an emerald green colour due to the presence of the quinoline group. This test can be used in the identification of quinine alkaloids.
Rosequin test	Dilute acetic acid and a few drops of bromine water are added with a sample of alkaloids. Then a drop of a solution of potassium ferrocyanide and a drop of strong ammonia solution is added to the sample solution. In this solution, a few ml of chloroform is added then the chloroform layer becomes red. This test is also known as the Erythroquinine test. This test is used in the identification of quinine alkaloids.

Test name	Method
Colour reactions	
Murexide test	Add potassium chlorate and dil. HCl in caffeine sample and evaporate to dry. It produces a red colour. Further, exposing it to strong ammonia vapour, the red colour converts into purple or violet colour. When caffeine reacts with HCl, it forms tetramethylalloxanthine which further in presence of ammonia forms ammonium salt of tetramethylpurpuric acid (murexide). This test is applicable for purine alkaloids, mainly caffeine.
Precipitation reactions	
Dragendorff's reagent test	Dragendorff's reagent is added to the alkaloid sample which gives a reddish-brown precipitate. Reagent composition: Bismuth nitrate: 8.00 g, Nitric acid: 21.5 g, Potassium iodide: 27.2 g, and water to make 100 ml.
Mayer reagent test	Mayer reagent is added to the alkaloid sample which gives a cream colour precipitate. Reagent composition: Mercuric chloride: 1.3 g, Potassium iodide: 5.00 g, and water to make 100 ml.
Wagner reagent test	Wagner reagent is added to the alkaloid sample which gives a brown colour precipitate. Reagent composition: Iodine: 1.3 g, Potassium iodide: 2 g, and water to make 100 ml.
Hager reagent test	Hager reagent is added to the alkaloid sample which gives a yellow colour precipitate. Reagent Composition: Picric acid: 1.0 g, water to make 100 ml.
Valser's test	Alkaloids react with Mercuric iodide giving a white colour precipitate.
Other tests	Alkaloid sample reacts with acids giving a buff colour precipitate and with picolinic acid they give a yellow colour precipitate.

4. Techniques of Extraction and Purification

Extraction is the first step to taking out the compounds from the source with the help of a suitable solvent (Fabricant and Farnsworth, 2001; Huie, 2002). Extraction is also the preliminary step in the drug discovery process and development. There is a wide range of well-established techniques to extract alkaloids from a source which include maceration, infusion, decoction, boiling under reflux, microwave-assisted extraction, ultrasound-assisted extraction, supercritical fluid extraction, and pressurized liquid extraction (Table 9). Because of their vast array of pharmacological actions, most of the alkaloids are purified from the crude extract by acid-base extraction (Ziegler and Facchini, 2008). The technique of extraction and purification of alkaloids is shown in Figure 8.

Table 9. Different types of extraction processes for alkaloids

Type	Temperature	Pressure/Time	Investment	Inference
Soxhlet	Depends upon the solvent	Atmospheric/ 6–8 h	Very low	Efficient for polar and nonpolar compounds
Supercritical fluid extraction	–	25–45 MPa; 1–2 h	High	Efficient for nonpolar compound
Pressurized liquid extraction	80–200°C	1–10 MPa; 10–30 min	High	Efficient for polar and nonpolar compounds
Microwave-assisted extraction	80–150°C	Variable/ 10–30 min	Moderate	The use of solvents is risky
Hydro distillation	40–60°C	Atmospheric/ 6–8 h	Low	Residual hydrotropic
Maceration	Room temperature	Atmospheric/ 7 days	Low	Used for hard drugs
Infusion	Room temperature	Atmospheric/ variable	Low	Useful for soft drugs
Decoction	At boiling temperature	Atmospheric/ until remains half	Low	Not used for thermolabile substances
Boiling under reflux	At boiling temperature	Atmospheric/ variable	Moderate	Efficient technique

4.1. Mechanical Methods of Extraction

Mortar-pestle, mixer grinder and grinding mill are the common mechanical methods for grounding plant tissues before extraction. In case of instant need extraction, the lignified plant material was frozen and pulverized with liquid nitrogen in a mortar pestle (Kirakosyan et al., 2016). In some literature, a sonicator is also used for the extraction of plant products grown by cell suspension culture methods (Bhargavi et al., 2018).

4.1.1. Precautions to Be Taken During Extraction Processes

Alkaloids may be degraded in presence of excessive heat, certain chemicals, and enzymatic reactions inside the cell; hence cautions should be taken respectively. In the case of water-soluble materials, the pH should be taken into consideration and a buffer system can be utilized for such purpose. Due to methodical error either in the extraction technique or in the isolation process, polyphenols may be deactivated by insoluble polyvinyl polypyrrolidine, and soluble polyvinyl pyrrolidine, so these contraindicated chemicals should not be used together. A chelating agent, commonly EDTA may be added to eliminate divalent cations from the extracts. Antifoaming agents such as Dow Corning or Union Carbide can be used to suppress foam if produced during the extraction process (Bhargavi et al., 2018; Fabricant and Farnsworth, 2001; Huie, 2002).

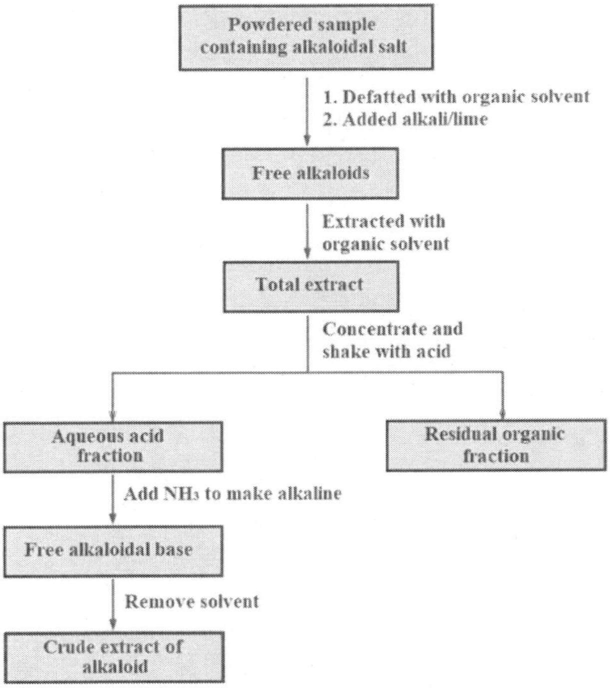

Figure 8. General flow chart of alkaloid extraction and purification.

4.2. Extraction with Solvents

Extraction is a separation technique of the desired compound from a matrix in which soxhlet extraction and counter-current extraction methods are generally utilized (Bhargavi et al., 2018; Fabricant and Farnsworth, 2001; Huie, 2002; Pandey and Tripathi, 2014). The extraction with solvent may be categorised as liquid-liquid extraction or solid-liquid extraction. In this technique, the homogenized cell is kept in the extracting solvent for a certain time so that all parts of the cells are easily penetrated by the solvent. The main primary key of the extraction process is to find out the suitable solvent system and temperature effect. An ideal extraction method should avoid some important phenomena associated with physical incompatibilities like decomposition, isomerization, or polymerization. Another important parameter is compound stability, which depends on light, heat, and solvent polarity.

An ideal solvent dissolves only those compounds which are required by leaving the other constituents. The strength of the solvent interaction with various polar solutes can be measured by the polarity index. For extraction of the nonpolar alkaloids, nonpolar solvents are used (Barnes, 1999; Pandey and Tripathi, 2014; Patil et al., 2012). The volatility of solvent is a major problem that occurs during the extraction process, which may be overcome by using solvents with a low boiling point without denaturation at high temperatures. Solvent selection is indeed the key to better extraction but if the choice of solvent is not available then it can be modified and by modification, the solvent becomes more acidic and more basic in certain extraction processes.

For the extraction of alkaloids, acidified solvents are mainly used, whereas the basic natures of certain solvents are utilized for the extraction of phenolic compounds. It has been observed from the literature that 5% HCl is best suited for alkaloid extraction (Barnes, 1999; Magistretti, 1980; Pandey and Tripathi, 2014).

5. Classifications of Extraction Processes

Based on heat the extraction can be classified as hot and cold extractions. Depending upon the physical state, the extraction can be classified as solid-liquid extraction or liquid-liquid extraction (Cos et al., 2006; Culture and Health, 1996; Pandey and Tripathi, 2014; Barnes, 1999). However, based on the protocol, these extraction processes are further sub-classified and are summarized in this section.

5.1. Hot and Cold Extraction

Hot extraction is the most commonly used process which is associated with prolonged time and higher temperature. Hot extraction processes include decoction, digestion, reflection, steam distillation etc. However, the excess temperature in the hot extraction method may polymerize or decompose the thermolabile compounds (Azwanida, 2015; Barnes, 1999). To avoid these disadvantages, the cold extraction process is preferred and this is performed at room temperature. Several types of cold extraction techniques are available for small as well as large-scale purposes like percolation, maceration, superfluid extraction etc. (Harborne, 1984).

5.2. Liquid-Liquid and Solid-Liquid Extractions

The liquid-liquid method is applied if the compound is liquid, and extraction depends on the partition coefficient. The distribution coefficient between the liquid form and solvent is appreciable in this case. On the other hand, if the precursor is solid, the solid-liquid extraction is carried out (Farhat et al., 2009; Ozek et al., 2010; Sahraoui et al., 2008). As per the protocol adopted in the case of extraction from solids, there are several subtypes of solid-liquid extraction techniques which are described as follows.

5.2.1. Maceration

In this method, a crude drug is first cuts into small pieces and then transferred in a stoppered container and covered with a solvent for a certain time until the solubilized part is dissolved in the solvent. In general, the extraction takes place in maceration is 7 days and is applicable for hard drugs. It is an example of the cold extraction process (Cunha et al., 2004; Majors, 1996; Pandey and Tripathi, 2014; Phrompittayarat et al., 2007; Ronald, 1999; Sasidharan et al., 2008).

5.2.2. Percolation

It is suitable for the extraction of a soft kind of compound and is carried out within 24 hr. Plant products are transferred in a percolation tube plugged with cotton with a filter and then dipped in the solvent. The total experiment is carried on at room temperature. The extract is then collected by a stopper below (Figure 9) (Saberi, 2015).

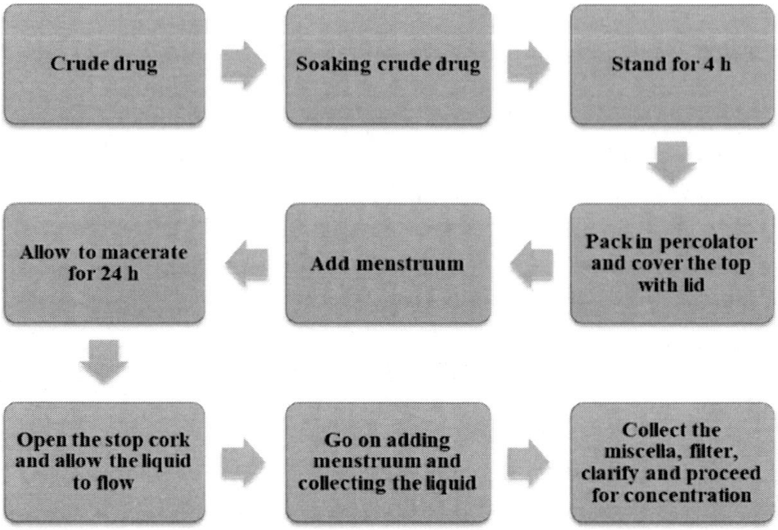

Figure 9. Schematic representation for the extraction using the percolation procedure. (Partially adapted from Mukherjee, 2019).

5.2.3. Digestion

In this method, a temperature of about (40–60°C) is applied at the time of the extraction process. The process is suitable for thermostable plant materials. Modification of the method is done by mixing the plants' products using a magnetic stirrer and a mechanical stirrer. The extract is filtered after 8–12 h, and then the fresh solvent is added. The method is repeated until the extraction of desired solutes (Komárek et al., 2006; Pandey and Tripathi, 2014).

5.2.4. Infusion

Infusion is the technique in which, the plant material is extracted by cold water or hot water for a short time. By allowing the leaves, flowers, soft stems, citrus peeling etc to steep in the water, the compound leach into the water (Pandey and Tripathi, 2014).

5.2.5. Decoction

The decoction is a process in which source material is boiled for a specified time i.e., until the solvent remains half of its initial amount in the assembly.

The process is appropriate for the extraction of thermostable and water-soluble plant materials (Pandey and Tripathi, 2014).

5.3. Extraction with Reflexion

In this technique of extraction, the plant material is kept with water in a flask. The heat is applied to boil the solvent and the vapours move upward were a closed assembly with a condenser fitted on top. The solvent vapours are condensed by the condenser and recycled. The process remains to continue until complete extraction occurs (Johansen et al., 1996; Mahajan et al., 2015).

5.4. Pressurized Liquid Extraction

This extraction method is also known as accelerated solvent extraction or enhanced solvent extraction. In this technique, temperature and pressure gradually increase where an increase in temperature promotes the extraction process by increasing the diffusivity of the solvent, whereas an increase in pressure can promote the penetration process through matrix pores without altering the liquid state of the organic solvent. The technique provides better working efficiency under inert atmospheric conditions and with protection from light. This extraction process requires less solvent and takes less time (Bjorn et al., 1999; Liu et al., 2013; Richter et al., 1996; Skalicka-Wozniak and Glowniak, 2012).

5.5. Soxhlet Extraction

Soxhlet extraction is the technique of continuous extraction in which plant material is continuously exposed to a hot solvent. Soxhlet apparatus is constructed with unique glass materials used for organic solvent extractions purpose and borosilicate glass is always used for such purposes. During the procedure of extraction, the solvent is first boiled gently inside a round bottom flask, and then the vapour passes through the side tube, where it is condensed by a condenser, and falls drop by drop into thimble-containing material. In this way, Soxhlet is filled up gradually, and when it reaches the top of the tube it siphons over into the flask. The process keeps continues till the colourless solvent does not collect (Zygmunt and Namieśnik, 2003).

5.6. Supercritical Fluid Extraction

Supercritical fluids such as CO_2, ethylene, ethane, propylene, propane, and nitrous oxide, are the most popular but CO_2 has a comparatively low critical temperature (31.1°C) and pressure (73.8 bar). The method has been used for large-scale extraction and is also suitable for the extraction of thermolabile compounds (Lutfun and Sarker, 2012; Patil and Shettigar, 2010; Zougagh et al., 2004).

5.7. Ultrasonic Extraction

In this technique, the plant tissues are exposed to high-frequency sound to liberate the phytochemicals. The ultrasound effect promotes the rate of extraction when it is used with mixtures of immiscible solvents. The limitation of this technique is that it generates heat, which affects the extraction of thermolabile products. Hence, to avoid such types of problems, extraction is carried out under reduced temperature in an ice bath (Vinatoru, 2001).

5.8. Microwave-Assisted Extraction

Electromagnetic radiations are used in microwave-assisted extraction, in which a microwave frequency of about 2.45 GHz is used for general domestic extraction purposes. In this technique, the microwave energy is passed through the solvent, with brief periods of cooling time as the process generates much heat, where the electric field is responsible for the heating of substrates through the dipolar rotation and ionic conduction. During the extraction process, the temperature is increased gradually due to a higher dielectric constant which breaks down the weak hydrogen bonds. Hence, the extractions of thermolabile compounds with a low dielectric constant are not extracted in normal environmental conditions and need a cold environment (Delazar et al., 2012).

5.9. Solid-Phase Extraction

The solid-phase extraction technique needs cartridges and disks with a variety of sorbents, where the solute molecules are attached over the surfaces of the stationary phase. Normal phase, reverse phase, and ion exchange solid-phase extraction units are available for the extraction of different kinds of alkaloids (Tekel and Hatrík, 1996; Buchholz and Pawliszyn, 1994; Zini et al., 2001).

5.10. Sequential and selective extraction

In sequential extraction, also known as successive extraction, the extraction is carried out on the same plant material successively in the order of polarity of the solvent whereas, in selective extraction, a particular solvent is used for the extraction and once the extraction is over, the fresh plant material is used for further extraction with other solvents (Vankar, 2004).

6. Drying of Extract

6.1. Application of Heat

The heating bath temperature is preferred over others in which around 20–30°C higher temperature is applied than the boiling point of the solvent. If the boiling point of the solvent is below 80°C, then the extract is concentrated in a water bath, whereas a heating mantle, sand bath, or hot plate is used for solvents having a higher boiling point.

6.2. Rotary Evaporator

The technique is mainly used for the drying of a thermolabile product where the solvent is removed by using a vacuum evaporator. It works on the principle of the PV ¼ nRT equation, which shows the directly proportional relationship between pressure and temperature, thus solvents boil at reduced temperature. In the rotary evaporator, water boils at 40°C at a vacuum of 72 mbar, ethanol boils at 40°C at a vacuum of 240 mbar, and butanol boils at 40°C at 25 mbar. Molecular weight, density, and vapour are the important factors which affect the boiling of a liquid. The rotation is applicable for the proper heat transfer with a reduction of the bumping phenomenon of solvents. The rotation also forms a layer inside the flask and increases the surface area, which ultimately increases the rate of evaporation. For the evaporation of low boiling solvents, acetone and dry ice are used as coolants, whereas, methanol, ethanol, or isopropanol is used as anti-freeze agents when very low temperature is required (Barnes, 1999; Kahol et al., 1998; Pandey and Tripathi, 2014; Wang and Weller, 2006).

6.3. Freeze Drying

Freeze drying is mainly used for the drying of thermolabile substances. It works on the principle of sublimation where the aqueous solution is frozen and the ice is sublimed off to leave a dry residue. The number of alkaloids can be dried by freeze drying method in which the material is initially cooled below its triple point. The drying is completed in two stages i.e., primary drying stage and secondary drying stage. In the primary drying stage, about 95% of the water in the compound is sublimated and the sublimed water vapour becomes condensed. In the secondary drying stage, the residual water molecules are removed from the frozen material and the residual water content reaches around 0.5% (Barnes, 1999; Pandey and Tripathi, 2014; Wang and Weller, 2006).

7. Pharmaceutical Applications of Alkaloids

Some of the renowned characters from history are associated with the use of alkaloid-based extracts, such as Socrates, a great philosopher, was passed away due to the use of *Conium maculatum* extract and Cleopatra was known for using *Hyoscyamus muticus* to dilate her pupils to achieve a more attractive appearance. In Medieval Europe, *Atropa belladonna* was frequently used by women aiming for the same results as Cleopatra, being the alkaloid, coniine, the responsible compound in this particular case. Later on, in history, its derivatives began to be used during medical examinations for dilating pupils. Tropicamide is another similar example since it has been used for Alzheimer’s disease diagnosis (Ncube et al., 2015).

The pharmaceutical applications of alkaloids are very wide. For example, strychnine is used for many purposes like the treatment of multiple disorders and eye conditions. The active component of Dysurgal or Pasuma is strychnine use in clinical doses (Aniszewski, 2007; Jeffrey et al., 1994; Smeller and Wink, 1998).

In recent days alkaloids are very common components of herbal medicines. Researchers are chemically modifying the structure of alkaloids for a better therapeutic response as after modifications, they show a better response than natural drugs. However, natural alkaloids are equally significant in phototherapy, homoeopathy, and alternative medicine (Aniszewski, 2007; Sayeed et al., 2013).

Some alkaloids have clinical importance such as indole, isoquinoline, and tropane derivatives. Tropane derivatives like atropine, hyoscine, and hyoscyamine are promoted by pharmaceutical companies on a large scale and also for clinical purposes. There are more than fifty different alkaloid-based products that have been developed and marketed. Some popular marketed alkaloidal products are atropine sulphate from atropinol, hyoscine derivative buscopan which is used in transdermal plasters and hyoscyamine from bella sanol (Aniszewski, 2007; Drager, 2008; Grzegorz and Gadzikowska, 2008).

Important alkaloids such as boldine, codeine, narceine, morphine etc. have a significant role in clinical therapy. Oxyboldine alkaloid has morphine-like pharmacological action. Codeine is one of the most popular alkaloidal compounds which is present in more than 250 pharmaceutical preparations on the market. Codicaps or Codipront can be utilized for clinical purposes. Most of the alkaloidal products are modified from opium i.e., narceine-containing drugs are related to codeine and are mainly used for cough treatment (Aniszewski, 2007; Sayeed et al., 2013).

Tubocurarine derivatives like tubarine or jexin are used as a muscle relaxant. Morphine-containing drugs such as morphalgin and spasmofen are used in serious cases like surgical operations and postoperation treatments (Aniszewski, 2007).

Chemical constituents of indole alkaloids such as ephedrine, ergotamine, ergometrine, and yohimbine are used in different formulations alone or in combination (Kumar et al., 2013a). The main active ingredient of Dorex or Endrine is ephedrine which is used in the treatment of nasal cold symptoms or bronchial asthma (Aniszewski, 2007). The main chemical constituent of ergot is ergotamine which is extremely available in the market due to its wide applications. Ergostat or Migral have marketed products based on ergotamine and are generally used for treating migraines. Yohimbine is the main active molecule of aphrodyne or yohimex and based on yohimbine approx 20 different compounds have been developed, which are utilized in impotency-related problems in men (Aniszewski, 2007; Jeffrey et al., 1994; Kumar et al., 2013a; Smeller and Wink, 1998). In modern science, alkaloids are utilised as anaesthetics, stimulants, antibacterials, antimalarials, analgesics, antihypertensive agents, spasmolysis agents, anticancer agents, antiasthmatic preparations, vasodilators, antiarrhythmic agents, etc. These properties, as well as their toxicity, continue to be an important research field (Kuete, 2014). Pharmaceutical applications of some alkaloids are enlisted in Table 10.

Table 10. Pharmaceutical applications of some important alkaloids

Alkaloid	Pharmaceutical Application
Physostigmine	Acetylcholinesterase inhibitor
Morphine	Analgesic
Quinidine, Ajmaline	Antiarrhythmic
Ephedrine	Antiasthma
Chelerythrine	Antibacterial
Homoharringtonine	Anticancer
Piperine	Antihyperglycemic activities
Reserpine, Vincamine	Antihypertensive
Emetine	Antiprotozoal agent
Quinine	Antipyretic, antimalarial
Vinblastine, Vincristine	Antitumor
Glaucine	Antitussive
Yohimbine	Aphrodisiac
Galantamine	Cholinomimetic
Tubocurarine	Muscle relaxant
Psilocin	Psychotropic
Cocaine, Caffeine, Nicotine, Theobromine	Stimulant
Ergot alkaloids	Vasoconstriction, Hallucinogenic, Uterotonic
Vincamine	Vasodilatory

Conclusion

Alkaloids acquire an exceptional therapeutic and public interest as a source of novel lead compounds for drug development against various disease ailments. In the current scenario, there are so many alkaloids are utilizing as drugs, against their poisonous effects, i.e., caffeine as a psychostimulant, codeine as an antitussive, cocaine as a local anaesthetic, morphine as an indispensable analgesic etc. Today, alkaloids are clinically applied for the treatment of a wide range of diseases including cancer, diabetes, and neurological disorders. Researchers are continuously exploring new and new pharmaceutical applications of alkaloids for the welfare of mankind. Still, there are a lot of scopes to explore more of the pharmacological benefits of alkaloids in humans. With the help of the structure-activity relationship, the safety and efficacy of existing alkaloids can be improved. Furthermore, the metabolic transformation of alkaloids, the introduction of a wide range of scientific tools, and the finding of novel sources of alkaloids are still opportunities for researchers in near future.

References

Alves de Almeida, A. C., de-Faria, F. M., Dunder, R. J., LPB, M., ARM, S. B., Luiz-Ferreira, A., (2017). Recent trends in pharmacological activity of alkaloids in animal colitis: potential use for inflammatory bowel disease. *Evid Based Complement Alternat Med*, 2017, 8528210.

Amirkia, V., Heinrich, M., (2014). Alkaloids as drug leads–a predictive structural and biodiversity based analysis. *Phytochem Lett*, 10, xlviii–xlliii.

Aniszewski, T., (2015). *Alkaloids: chemistry, biology, ecology, and applications*. Elsevier, Pacific Grove, CA, p. 496.

Aniszewski, T. (2007). *Alkaloids—Secrets of Life. Alkaloid Chemistry, Biological Significance, Applications and Ecological Role*. Elsevier.

Ashihara, H., Yokota, T., Crozier, A., (2013). Purine Alkaloids, Cytokinins, and Purine-Like Neurotoxin Alkaloids. *Natural Products*, Springer, pp. 953–975.

Ata, A., Andersh, B. J., (2008). Buxus steroidal alkaloids: chemistry and biology. *Alkaloids. Chem. Biol.*, 66, 191–213.

Azwanida, N., (2015). A review on the extraction methods use in medicinal plants, principle, strength and limitation. *Med. Aromat. Plants*, 4, 3–8.

Baker, B., (1996). *Alkaloids: Chem. Biol. Perspect.* Elsevier, 1(ed), 10, 357–407.

Barnes, J., (1999). Trease and Evans' pharmacognosy. *Focus. Altern. Complement. Ther.*, 4, 151–152.

Bauer, I, Knolker H. J., (2012). Synthesis of pyrrole and carbazole alkaloids. In: Knolker H. J (ed) *Alkaloid synthesis*. Springer-Verlag, Berlin Heidelberg, pp. 203–253.

Bentley, K. W., (2006). β-Phenylethylamines and the isoquinoline alkaloids. *Nat. Prod. Rep.*, 23, 444–463.

Beyer, J., Drummer, O. H., Maurer, H. H., (2009). Analysis of toxic alkaloids in body samples. *Forensic Sci Int.*, 185(1–3), 1–9.

Bhargavi, G., Rao, P. N., Renganathan, S., (2018). Review on the Extraction Methods of Crude Oil From All Generation Biofuels in Last Few Decades, IOP Conference Series: Materials Science and Engineering. IOP Publishing, p. 012024.

Biastoff, S., Drager, B., (2007). *Calystegines. Alkaloids: Chem. Biol.*, 64, 49–102.

Bjorn, B., Danz, H., Hamburger, M., (1999). Pressurized liquid extraction of medicinal plants. *J. Chromatogr. A.*, 837, 211–219.

Bottger, A., Vothknecht, U., Bolle, C., Alkaloids, W. A. (2018). Lessons on Caffeine, Cannabis & Co. Springer, Berlin Heidelberg, pp. 179–203.

Bribi, N., (2018). Pharmacological activity of alkaloids: a review. *Asian J. Bot.,* 1. https://doi. org/10.63019/ajb.v1i2.467.

Bribi, N., Belmouhoub, M., Maiza, F., (2017). Analgesic and anti-inflammatory activities of ethanolic extract of fumaria capreolata. *Phytothérapie,* 15(4), 211-216.

Bruneton, J., (1999). *Pharmacognosy, phytochemistry, medicinal plants*. Lavoisier, 2nd Edition, p. 1136.

Buchholz, K., Pawliszyn, J., (1994). Optimization of solid-phase microextraction conditions for determination of phenols. *Anal. Chem.*, 66, 160–167.

Bunsupa, S., Yamazaki, M., Saito, K. (2012). Quinolizidine alkaloid biosynthesis: recent advances and future prospects. *Front Plant Sci.,* 3, 239.

Cheeke, P., (1988). Toxicity and metabolism of pyrrolizidine alkaloids. *J. Anim. Sci.,* 66, 2343–2350.

Chen, J., Gao, K., Liu, T., Zhao, H., Wang, J., Wu, H., Liu, B., Wang, W. (2013). Aporphine alkaloids: a kind of alkaloids' extract source, chemical constitution and pharmacological actions in different botany. *Asian J. Chem.*, 25, 18.

Cordell, G. A. (1991). *The alkaloids. Chemistry and pharmacology*, Vols 41. Academic Press, London.

Cos, P., Vlietinck, A. J., Berghe, D. V., Maes, L., (2006). Anti-infective potential of natural products: how to develop a stronger *in vitro* 'proof-of-concept'. *J. Ethnopharmacol.*, 106, 290–302.

Croteau, R., Kutchan, T. M., Lewis, N.G., (2000). Natural products (secondary metabolites). *Biochem. Mol. Biol. Plants.*, 24, 1250–1319.

Culture and Health, (1996). *Orientation Texts-World Decade for Cultural Development* 1988–1997. Document CLT/ DEC/PRO-1996, p. 129.

Cunha, I., Sawaya, A., Caetano, F., Shimizu, M., Marcucci, M., Drezza, F., Povia, G., Carvalho, P., (2004). Factors that influence the yield and composition of Brazilian propolis extracts. *J. Braz. Chem. Soc.*, 15, 964–970.

Delazar, A., Nahar, L., Hamedeyazdan, S., Sarker, S., (2012). Microwave-assisted extraction in natural products isolation. In: *Methods Mol. Bio*l. Humana Press, pp. 89–115.

Dey, P., Kundu, A., Chakraborty, H. J., Kar, B., Choi, W. S., Lee, B. M., Bhakta, T., Atanasov, A. G., Kim, H. S., (2018). Therapeutic value of steroidal alkaloids in cancer: current trends and future perspectives. *Int. J. Cancer*, 145, 1731–1744.

Diaz, G., (2015). Toxicosis by plant alkaloids in humans and animals in Colombia. *Toxins* (Basel), 7(12), 5408–5416.

Dostal, J., (2000). Two faces of alkaloids. *J Chem Educ.*, 77(8), 993.

Drager, B., (2008). Biotechnology of Solanaceae alkaloids: a model or an industrial perspective? In: Kayser, O., Quax, W. J. (Eds.), *Medicinal Plant Biotechnology: From Basic Research to Industrial Applications.* Wiley-VCHVerlag GmbH & Co. KGaA, pp. 237–265.

Eguchi, R., Ono, N., Hirai Morita, A., (2019). Classification of alkaloids according to the starting substances of their biosynthetic pathways using graph convolutional neural networks. *BMC Bioinformatics*, 20, 380.

Encycloaedia Britannica, (2018). *Alkaloid.* https://www.britannica.com/science/alkaloid.

Estévez, V., Villacampa, M., Menéndez, J. C., (2014). Recent advances in the synthesis of pyrroles by multicomponent reactions. *Chem. Soc. Rev.*, 43(13), 4633–4657.

Evans, W. C., (2009). *Trease and evans' pharmacognosy.* Saunders Ltd., Elsevier, Edinburgh, pp. 353-415, 616.

Fabricant, D. S., Farnsworth, N. R., (2001). The value of plants used in traditional medicine for drug discovery. *Environ. Health Perspect*, 109, 69.

Facchini, P. J., (2001). Alkaloid biosynthesis in plants: biochemistry, cell biology, molecular regulation, and metabolic engineering applications. *Annu. Rev. Plant Physiol. Plant Mol. Biol.*, 52, 29–66.

Farhat, A., Ginies, C., Romdhane, M., Chemat, F., (2009). Eco-friendly and cleaner process for isolation of essential oil using microwave energy: experimental and theoretical study. *J. Chromatogr. A.*, 1216, 5077–5085.

Farias, F. M., (2006). Universidade Federal do Rio Grande do Sul; Porto Alegre, RS, Brazil: Psychotria myriantha mull arg. (rubiaceae): Caracterizacao dos alcalóides e avaliacao das atividades antiquimiotáxica e sobre o sistema nervoso central. [Federal University of Rio Grande do Sul; Porto Alegre, RS, Brazil: Psychotria myriantha mull arg. (rubiaceae): *Characterization of alkaloids and evaluation of antichemotaxica activities and on the central nervous system.*]

Fresneda, P. M., Molina, P., (2004). Application of iminophosphorane-based methodologies for the synthesis of natural products. *Synlett.*, 2004(1), 1–17.

Gao, F., Li, Y. Y., Wang, D., Huang, X., Liu, Q., (2012). Diterpenoid alkaloids from the Chinese traditional herbal "Fuzi" and their cytotoxic activity. *Molecules,* 17(5), 5187–5194.

Gossauer, A., (2003). Monopyrrolic Natural Compounds Including Tetramic Acid Derivatives, Fortschritte der Chemie organischer Naturstoffe [*Progress in the Chemistry of Organic Natural Products*]. Springer, pp. 1–188.

Goyal, S., (2013). *Ecological role of alkaloids.* Natural Products: Phytochemistry, Botany and Metabolism of Alkaloids, Phenolics and Terpenes, 149–71.

Grzegorz, G., Gadzikowska, M., (2008). Tropane alkaloids as medicinally useful natural products and their synthetic derivatives as new drugs. *Pharmacol. Rep.*, 60, 439.

Gutierrez-Grijalva, E. P., Lopez-Martinez, L. X., Contreras-Angulo, L. A., Elizalde-Romero, C. A., Heredia, J. B., (2020). Plant Alkaloids: Structures and Bioactive Properties. In: *Plant-derived Bioactives*, M. K. Swamy (ed.). Springer Nature Singapore, pp. 85–117.

Habermehl, G., (1967). The Steroid Alkaloids: The Salamandra Group, *The Alkaloids: Chemistry and Physiology.* Elsevier, pp. 427–439.

Hegnauer, R., (1963). *The Taxonomic Significance of Alkaloids.* 14. doi:10.1016/B978-0-12-395540-1.50018-XCorpus ID: 89030452.

Hemscheidt, T., (2000). *Tropane and Related Alkaloids, Biosynthesis.* Springer, pp. 175–206.

Henning, C. P. (2013). Compuestos secundarios nitrogenados: alcaloides. [Nitrogenous secondary compounds: alkaloids.] In: Ringuelet J, Vina S (eds) *Productos Naturales Vegetales.* Editorial de la Universidad de la Plata, [Natural Vegetable Products. Editorial of the University of La Plata], La Plata, Argentina, p. 18.

Huie, C., (2002). A review of modern sample-preparation techniques for the extraction and analysis of medicinal plants. *Anal. Bioanal. Chem.*, 373, 23–30.

Hussain, G., Rasul, A., Anwar, H., Aziz, N., Razzaq, A., Wei, W., Ali, M., Li, J., Li, X., (2018). Role of plant derived alkaloids and their mechanism in neurodegenerative disorders. *Int. J. Biol. Sci.,* 14(3), 341.

Jayakumar, K., Murugan, K., (2016). Solanum alkaloids and their pharmaceutical roles: a review. *J, Anal. Pharm. Res.*, 3(6), 00075.

Jeffrey, H., Baxter, H., Webster, F., (1994). Phytochemical dictionary: a handbook of bioactive compounds from plants. *J. Chem. Ecol.,* 20, 815–818.

Jerzykiewicz, J., (2007). Alkaloids of Solanaceae (nightshade plants). *Postepy Biochemii,* 53(3), 280-286.

Jin, Z., (2006). Imidazole, oxazole and thiazole alkaloids. *Nat. Prod. Rep.,* 23, 464–496.

Jing, H., Liu, J., Liu, H., Xin, H., (2014). Histochemical investigation and kinds of alkaloids in leaves of different developmental stages in Thymus quinquecostatus. *Sci. World J.,* 2014, 839548.

Johansen, H., Glitso, V., Knudsen, K., (1996). Influence of extraction solvent and temperature on the quantitative determination of oligosaccharides from plant materials by high-performance liquid chromatography. *J. Agric. Food Chem.*, 44, 1470–1474.

Kahol, A., Tandon, S., Singh, K., (1998). Developments in the separation technologies for the processing of medicinal and aromatic plants. *Chem. Week. Bombay*, 44, 145–154.

Kam, T., Choo, Y., (2006). Bisindole alkaloids. *Alkaloids: Chem. Biol.*, 63, 181–337.

Kaur, R., Arora, S., (2015). Alkaloids-important therapeutic secondary metabolites of plant origin. *J. Crit. Rev.,* 2(3):1–8.

Kaur, R., Matta, T., Kaur H. (2019). Plant derived alkaloids. *Saudi J. Life Sci.* 2(5):158–189.

Kawasaki, T., Higuchi, K., (2005). Simple indole alkaloids and those with a nonrearranged monoterpenoid unit. *Nat. Prod. Rep.*, 22, 761–793.

Keeler, R. F., (1986). Teratology of steroidal alkaloids. *Alkaloids: Chem. Biol. Perspect,* 4, 389–425.

Kirakosyan, A., Noon, K. R., McKenzie, M., Cseke, L. J., Kaufman, P. B., (2016). Isolation and Purification of Metabolites, *Handbook of Molecular and Cellular Methods in Biology and Medicine*. CRC Press, pp. 386–399.

Knolker, H. J., (2008). Synthesis of biologically active carbazole alkaloids using selective transition-metal-catalyzed coupling reactions. *Chem. Lett.*, 38, 8–13.

Knunyants I., Zefirov N., (1988). Sov. Entsikl.*;* Moscow: *Chemical Encyclopedia.*

Koleva, I. I., van Beek, T. A., Soffers, A. E., Dusemund, B., Rietjens, I. M., (2012). Alkaloids in the human food chain–natural occurrence and possible adverse effects. *Mol. Nutr. Food. Res.*, 56(1), 30–52.

Komárek, M., Chrastný, V., Ettler, V., Tlustos, P., (2006). Evaluation of extraction/digestion techniques used to determine lead isotopic composition in forest soils. *Anal. Bioanal. Chem.*, 385(6), 1109–1115.

Kuete, V. (2014). Health Effects of Alkaloids from African Medicinal Plants. In: Kuete V (ed) *Toxicological Survey of African Medicinal Plants*. Elsevier, London, pp. 611–633.

Kufe, D. W., Pollock, R. E., Weichselbaum, R. R., Bast, R. C., Gansler, T. S., Holland, J. F., (2003). *Holland-Frei cancer medicine*. 6th ed. Hamilton (ON), BC Decker Inc.

Kukula-Koch, W. A., Widelski, J. (2017). Alkaloids. In: Badal S, Delgoda R (eds) *Pharmacognosy*. Academic Press, Boston, pp. 163–198.

Kumar, K., Kaushik, N., Attri, P., Kumar, N., Kim, C., Verma, A., Choi, E., (2013a). Biomedical importance of indoles. *Molecules*, 18, 6620–6662.

Kutchan, T. M., (1995). Alkaloid biosynthesis—the basis for metabolic engineering of medicinal plants. *Plant Cell*, 7.

Ladenburg, A. (1881). Die Naturlich vorkommenden mydriatische wirkenden Alkaloide. [The naturally occurring mydriatic alkaloids.] *Ann. Chem. Pharm.* 206, 274-307.

Lean, M. E., Ashihara, H., Clifford, M. N., Crozier, A., (2011). Purine alkaloids: a focus on caffeine and related compounds in beverages. In: *Teas, Cocoa and Coffee: Plant Secondary Metabolites and Health*, pp. 25–44.

Lewis, R. A., (1998). CRC Press, *Lewis' Dictionary of Toxicology*.

Liu, A., Chao, H., Zhou, X., Zhu, Z., Huang, F., Shen, Y., (2013). Determination of three capsaicinoids in Capsicum annuum by pressurized liquid extraction combined with LC-MS/MS. *J. Sep. Sci.,* 36, 857–862.

Lovell, B. E., Butterfield, W. J. H., Mc Gehee, H. A., Heptinstall, R. H., Lewis, T., (1986). *International Dictionary of Medicine and Biology*. John Wiley & Sons, New York, USA, 14(1): 118.

Lu, J. J., Bao, J. L., Chen, X. P., Huang, M., Wang, Y. T. (2012). Alkaloids isolated from natural herbs as the anticancer agents. *Evid. Based Complement Alternat. Med.,* 2012.

Lutfun, N., Sarker, S., (2012). Supercritical fluid extraction in natural products analyses. In: John, W. (Ed.), *Methods in Molecular Biology*. Humana Press, pp. 43–74.

Maat, L., Beyerman, H., (1984). The Imidazole Alkaloids, *The Alkaloids: Chemistry and Pharmacology*. Elsevier.

Magistretti, M., (1980). Remarks on the pharmacological examination of plant extracts. *Fitoterapia,* 51, 67–79.

Mahajan, A., Shelke, U., Gadhave, P., Bhalerao, K., (2015). Essential oils: at a glance. *World J. Pharm. Pharm. Sci.,* 4, 718–738.

Mazid, M., Khan, T. A., Mohammad, F., (2011). Role of secondary metabolites in defense mechanisms of plants. *Biol. Med.,* 3(2), 232–249.

McNaught, A. D., (1997). Blackwell Science Oxford. *Compendium of Chemical Terminology*.

Meira, M., EPd, S., David, J. M., David, J. P. (2012). Review of the genus Ipomoea: traditional uses, chemistry and biological activities. *Revista Bras. Farmacog*., 22(3), 682–713.

Michael, J. P., (2008). Indolizidine and quinolizidine alkaloids. *Nat. Prod. Rep*., 25(1), 139–165.

Michael, J. P., 2008. Quinoline, quinazoline and acridone alkaloids. *Nat. Prod. Rep*., 25, 166–187.

Mukherjee, P. K., (2019). Chapter 6 - Extraction and Other Downstream Procedures for Evaluation of Herbal Drugs. *Quality Control and Evaluation of Herbal Drugs*, Elsevier, pp. 195-236.

Muthna, D., Cmielova, J., Tomsik, P., Rezacova, M., (2013). Boldine and related aporphines: from antioxidant. *Nat. Prod. Commun.*, 8(12), 1797-1800.

Ncube, B., Nair, J. J., Rarova, L., Strnad, M., Finnie, J. F., Van Staden, J., (2015). Seasonal pharmacological properties and alkaloid content in Cyrtanthus contractus. *N. E. Br. S. Afr. J. Bot.*, 97, 69–76.

O'Connor, S. E., (2010). *1.25-alkaloids Comprehensive Natural Products*. II 1, 977-1007.

Ozek, G., Demirci, F., Ozek, T., Tabanca, N., Wedge, D., Khan, S., Bas¸er, K. H. C., Duran, A., Hamzaoglu, E., (2010). Gas chromatographic-mass spectrometric analysis of volatiles obtained by four different techniques from Salvia rosifolia Sm., and evaluation for biological activity. *J. Chromatogr.,* A 1217, 741–748.

Pan, Q., Mustafa, N. R., Tang, K., Choi, Y. H., Verpoorte, R., (2016). Monoterpenoid indole alkaloids biosynthesis and its regulation in Catharanthus roseus: a literature review from genes to metabolites. *Phytochem. Rev.*, 15(2), 221–250.

Pandey, A., Tripathi, S., (2014). Concept of standardization, extraction and pre phytochemical screening strategies for herbal drug. *J. Pharmacogn. Phytochem.*, 2(5), 115-119.

Patil, P. S., Shettigar, R., (2010). An advancement of analytical techniques in herbal research. *J. Adv. Sci. Res.,* 1, 08–14.

Pelletier, S. W., (1983). The nature and definition of an alkaloid, in *Alkaloids: Chemical and Biological Perspectives*. Wiley, New York, USA, pp. 1-31.

Phrompittayarat, W., Putalun, W., Tanaka, H., Jetiyanon, K., Wittaya-areekul, S., Ingkaninan, K., (2007). Comparison of various extraction methods of Bacopa monnieri. *Naresuan Univ. J.: Sci. Technol.*, 15, 29–34.

Richter, B., Jones, B., Ezzell, J., Porter, N., (1996). Accelerated solvent extraction: a technique for sample preparation. *Anal. Chem.* 68, 1033–1039.

Ripperger, H., (1998). Solanum steroid alkaloids—an update. *Alkaloids: Chem. Biol. Perspect.,* 12, 103–185.

Guggisberg, A., Hesse, M., (2002). Alkaloids. In: *Encyclopedia of Physical Science and Technology*. Meyers R. A. (ed.). Academic Press, pp. 477-493.

Roberts, M. F., (2013). *Alkaloids: biochemistry, ecology, and medicinal applications*. Springer, New York.

Roberts, M. F., Wink, M., (1998). *Alkaloids: Biochemistry, ecology and medicinal applications*. Plenum Press, New York and London 62(4): 664.

Robertson, J., Stevens, K., (2014). Pyrrolizidine alkaloids. *Nat. Prod. Rep.,* 31, 1721–1788.

Robins, D. J., (1982). The pyrrolizidine alkaloids. In *Fortschritte der Chemie organischer Naturstoffe* [*Progress in the Chemistry of Organic Natural Products*], Springer Vienna., 115-202.

Ronald, M., (1999). *An overview of sample introduction and sample preparation of volatile organic compounds*. LC GC 17, S7–S13.

Rosemeyer, H., (2004). The chemodiversity of purine as a constituent of natural products. *Chem. Biodivers,* 1, 361–401.

Saberi, A. A., (2015). Recent advances in percolation theory and its applications. *Phys. Rep.,* 578, 1–32.

Sahraoui, N., Mavian, I., Bornard, C., Boutekedjiret, F. C., (2008). Improved microwave steam distillation apparatus for isolation of essential oils: comparison with conventional steam distillation. *J. Chromatogr. A.,* 1210, 229–233.

Sasidharan, S., Chen, Y., Saravanan, D., Sundram, K. M., Yoga Latha, L., (2011). Extraction, isolation and characterization of bioactive compounds from plants' extracts. *Afr. J. Tradit. Complement. Altern. Med*., 8, 1–10.

Sayeed, A., Garg, M., Tamboli, E., Abdin, M., Ansari, S., (2013). *In vitro* production of alkaloids: factors, approaches, challenges and prospects. *Pharmacogn. Rev*., 7, 27.

Schardl, C. L., Grossman, R. B., Nagabhyru, P., Faulkner, J. R., Mallik, U. P., (2007). Loline alkaloids: currencies of mutualism. *Phytochemistry*, 68, 980–996.

Schiff Jr., P. L., (1987). Bisbenzylisoquinoline alkaloids. *J. Nat. Prod*., 50, 529–599.

Schramm, S., Kohler, N., Rozhon, W. (2019). Pyrrolizidine alkaloids: biosynthesis, biological activities and occurrence in crop plants. *Molecules*, 24(3):498.

Silva Teles, M. M. R., Vieira Pinheiro, A. A., Da Silva Dias, C., Fechine Tavares, J., Barbosa Filho, J. M., Leitao Da Cunha, E. V., (2019). Alkaloids of the Lauraceae. In: Knolker H. J (ed) *The alkaloids: chemistry and biology*. Academic Press, San Diego, CA, pp. 147–304.

Skalicka-Wozniak, K., Glowniak, K., (2012). Pressurized liquid extraction of coumarins from fruits of Heracleum leskowii with application of solvents with different polarity under increasing temperature. *Molecules,* 17, 4133–4141.

Smeller, T., Wink, M., (1998). Alkaloids in modern medicine. In: Roberts, M., Wink, M. (Eds.), *Alkaloids. Biochemistry, Ecology, and Medicinal Applications.* Academic Press, New York, London, pp. 435–459.

Soares, M. S., Fernandes, J. B., Vieria, P. C., (2007). Alkyl, aryl, alkylarylquinoline, and related alkaloids. *Alkaloids: Chem. Biol.,* 64, 139–214.

Strunz, G. M., Findlay, J. A., (1985). Pyridine and piperidine alkaloids. In: *The Alkaloids: Chemistry and Pharmacology*. Elsevier, pp. 89–183.

Szoke, É., Lemberkovics, É., Kursinszki, L., (2013). Alkaloids derived from lysine: piperidine alkaloids. *Natural Products: Phytochemistry, Botany and Metabolism of Alkaloids, Phenolics and Terpenes*, 303–41.

Talapatra, S. K., Talapatra, B., (2015). Alkaloids. General Introduction. In: *Chemistry of Plant Natural Products: Stereochemistry, Conformation, Synthesis, Biology, and Medicine.* Springer Berlin Heidelberg, Berlin, Heidelberg, pp. 717–724.

Tekel, J., & Hatrík, S., (1996). Pesticide residue analyses in plant material by chromatographic methods: clean-up procedures and selective detectors. *J. Chromatogr*. A., 754, 397–410.

Tomko, J., Voticky´, Z., (1973). Steroid alkaloids: the Veratrum and Buxus groups. *Alkaloids: Chem. Physiol.,* 14, 1–79.

Vankar, P. S., (2004). Essential oils and fragrances from natural sources. *Resonance*, 9, 30–41.

Vinatoru, M., (2001). *An overview of the ultrasonically assisted extraction of bioactive principles from herbs*. Ultrason. Sonochem., 8, 303–313.

Wang, F. P., Chen, Q. H., Liu, X. Y., (2010). Diterpenoid alkaloids. *Nat. Prod. Rep.*, 27(4), 529–570.

Wang, L., Weller, C., (2006). Recent advances in extraction of nutraceuticals from plants. *Trends Food Sci. Technol.,* 17, 300–312.

Wansi, J. D., Devkota, K. P., Tshikalange, E., Kuete, V., (2013). Alkaloids from the medicinal plants of Africa. In: Kuete V (ed) *Medicinal Plant Research in Africa.* Elsevier, Oxford, pp. 557–605.

Wink, M., (ed), (2010). *Annual plant reviews, functions and biotechnology of plant secondary metabolites.* Blackwell Publishing Ltd, Annual Plant Reviews.

Wrobel, J. T., (1985). Pyrrolizidine alkaloids. *Alkaloids*, 26, 327–384.

Zhao, H., Wang, X. Y., Li, M. K., Hou, Z., Zhou, Y., Chen, Z., Meng, J. R., Luo, X. X., Tang, H. F., Xue, X. Y., (2015). A novel pregnane-type alkaloid from pachysandra terminalis inhibits methicillin-resistant Staphylococcus aureus *in vitro* and *in vivo. Phytother. Res.,* 29, 373–380.

Ziegler, J., Facchini, P. J., (2008). Alkaloid biosynthesis: metabolism and trafficking. *Annu. Rev. Plant Biol.,* 59, 735–769.

Zini, C. A., Augusto, F., Christensen, T. E., Smith, B. P., Carama˜o, E. B., Pawliszyn, J., (2001). Monitoring biogenic volatile compounds emitted by Eucalyptus citriodora using SPME. *Anal. Chem.*, 73, 4729–4735.

Zougagh, M., Valcárcel, M., Ríos, A., (2004). Supercritical fluid extraction: a critical review of its analytical usefulness. *Trends Anal. Chem.,* 23, 399–405.

Zygmunt, B., Namieśnik, J., (2003). Preparation of samples of plant material for chromatographic analysis. *J. Chromatogr. Sci.,* 41, 109–116.

Chapter 2

Tropane Alkaloids: Chemical Aspects and Recent Therapeutic Developments

Silky Sethy[1,*], Neha Minocha[2] and Rohini Agrawal[3]
[1]SGT College of Pharmacy, SGT University, Gurugram, Haryana, India
[2]School of Pharmacy, Chitkara University, Himachal Pradesh, India
[3]College of Pharmacy, JSS Academy of Technical Education, Noida, India

Abstract

Tropane alkaloids (TA) are treasured secondary plant metabolites that are frequently determined in high concentrations from Solanaceae and Erythroxylaceae family. The TAs, which can be characterized by their particular bicyclic tropane ring device, may be divided into 3 essential categories: hyoscyamine and scopolamine, cocaine and calystegines. Even though all TAs have the identical primary structure, they range immensely in their biological, chemical and pharmacological properties. Scopolamine, also known as hyoscine, has the most important significant marketplace as a pharmacological agent due to its role in treatment of nausea, vomiting, motion sickness, as well as smooth muscle spasms while cocaine is the second maximum regularly consumed illicit drug globally. This assessment presents a comprehensive review of TAs, highlighting their structural variations, use in pharmaceutical therapy from each ancient and modern perspective, herbal biosynthesis in plants and increasing production opportunities using tissue and microbial biosynthesis of those compounds.

Keywords: tropane alkaloids, solanaceae, cocaine, scopolamine, microbial biosynthesis

---

* Corresponding Author's Email: silkysethy1981@hotmail.com

In: The Essential Guide to Alkaloids
Editor: Deepak Kumar Semwal
ISBN: 979-8-88697-456-0

1. Introduction

Alkaloids are the compounds containing one or extra nitrogen atoms and name is derived from the fundamental nature of many individuals of this group, alkaloids from "alkaline-like." The definition of alkaloids is complicated as many nitrogen-containing molecules do now not necessarily belong to this group. Biogenic amines or amino sugars, as an example, are herbal plant products and N-containing however now not described as alkaloids. Tropane alkaloids (TAs) are selected class of alkaloid and defined as the molecules that possess a tropane ring (Barnes, 2000). TAs are either esters of 3α-tropanole (tropine) or to a lesser extent, three β-tropanole (pseudotropine) and can be divided into three groups. TAs from Solanaceae plants like hyoscyamine and scopolamine, coca alkaloids like cocaine from Erythoxylum coca and the lately calystegines group, which are polyhydroxylated nortropane alkaloids (NTAs) specially taking place in Convolvulaceae, Solanaceae, Moraceae, Erythrocylaceae and Brassicaceae (Bencharit et al., 2003).

In overall, ~2 hundred unique TAs were defined (Camps and Munoz-Torrero, 2002). Biosynthesis of the tropane ring system is homologous in organisms which produce these three TA. TA biosynthesis starts with the amino acid ornithine or arginine and their intermediate putrescine, continuing to the not unusual N-methyl-Δ^1-pyrrolinium cation precursor of all TAs. This is the branch point of cocaine, hyoscyamine/scopolamine and calystegine as well as nicotine biosynthesis (Carroll et al., 1999). Even though all TAs have excessive degree of structural similarity due to their tropane ring, the pharmacological effects of these compounds fluctuate drastically. Cocaine and hyoscyamine/scopolamine are able to bypass the blood-brain barrier and produce dose-structured hallucination and psychoactive outcomes. Calystegines do no longer cause those results because of their polarity as well as hydrophilicity and consequent incapacity to bypass blood-brain barrier.

As the calystegines are newly observed group of TAs without any pharmaceutical, medicinal or monetary interest, little research has to this point been performed in this group of TAs. In evaluation, the cultivation and production of scopolamine is of principal monetary interest due to its miscellaneous pharmaceutical packages. Certainly, global demand for this compound is increasing. Furthermore, scopolamine is one of the essential drugs of the sector fitness enterprise (Casale, 1987). Hyoscyamine and scopolamine are extracted from the Duboisia flowers cultivated on big plantations in Queensland, Australia (Barnes, 2000).

2. General Aspects

Tropane alkaloids are some of the oldest drugs associated with humans. Poisonous Solanaceae family plant life, Atropa, Brugmansia, Datura, Duboisia, Hyoscyamus, and Scopolia, which are today classed as genera and have several species that contain alkaloids, were well-known in the past and facts on their use in folk medicine by various ethnic businesses are abounding (Henry, 1949; Leake and Pelikan, 1976; Lounasmaa and Tamminen, 1993; Holzman, 1998; Griffin and Lin, 2000; Sneader, 2005). In 1819 Meissner (who without a doubt coined the term: alkaloid) comprehend that the active compounds of these poisonous plants are alkaline in nature and for this reason may be isolated by using extractive techniques and consequently alkaloid compounds started to be isolated from 1830 onward: atropine from *Atropa belladona* L, and hyoscyamine from *Hyoscyamus niger* L (Henry, 1949; Fodor, 1971; Sneader, 2005). Using the coca plant, which belongs to the Erythroxylaceae family, as a stimulant can further be traced to prehistoric times. Cocaine, the fundamental alkaloid of Erythroxylon coca was isolated in 1860. Ancient elements of the ethnopharmacological way of life and the start of medicinal chemistry, in which tropane alkaloid investigations performed a distinguished function, are well described in various sources (Fodor, 1971; Leake and Pelikan, 1976; Gyermek, 2002; Sneader, 2005). The modern-day medicinal drug makes use of lots of atropine and scopolamine extracted from genetically changed cultivars, whilst ever-developing demand complements new, chemical and biotechnological strategies in their production.

3. Tropanes of Plant Origin and their Traditional Medicinal Applications

Tropanes of plant origin and their conventional medicinal properties list many plants of a mild climate, whose extracts used for ages as poisons and health potions of diverse roles (e.g., anti-AChE, hallucinogens, and many others.) Plants contain extensive quantities of tropane alkaloids (e.g., up to two percent in ripe seeds of *Datura stramonium*), including lethal nightshade (*Atropa belladonna*), mandrake (*Atropa mandragora*), henbane (*Hyoscyamus niger*, *Hyoscyamus albus*), jimsonweed, also known as thornapple (*Datura stramonium*) and scopola (*Scopolia carniolica*), etc. (Griffin and Lin, 2000). Usually, all components of a plant incorporate alkaloids, which could lead through incidental consumption, to the poisoning of people or farm animals. With atropine as an example, death can arise after taking 100 mg, however cases of survival after 500 mg dose also are recorded. Because of its medicinal properties in modern times, those species have become a topic of collection (or eventually, cultivation) and ordinary exchange as a raw material for manufacturing of pharmaceutical industries isolate the pure active constituents, all through the 19th century (Fodor, 1971; Leake and Pelikan, 1976; Christen, 2000; Sneader, 2005). The name tropane is given to the bicyclic saturated shape (N-methyl-eight-azabicyclo[3.2.1]octane, characteristic of a category of ca. Two hundred alkaloids, are quite simply subdivided according to the number of carbons in the tropane skeleton and stereochemical functions. Despite the fact that tropine and its derivatives have a similar biogenetic reaction sequence that leads to the tropane skeleton, only the tropinone step is branched, their division in scientific literature has another crucial reason: the ecgonine significance is strongly diagnosed best with its predominant member – cocaine, a neurostimulator notorious for causing devastating addiction. The biosynthetic pathway on which the tropane derivatives are shaped has been thoroughly elucidated with the help of radioactive labeling experiments (Robins and Walton, 1993; Humphrey and O'Hagan, 2001; Patterson and O'Hagan, 2002; Oksman-Caldentey, 2007). In short, the principal precursor of the bicyclic alkamine element is L-ornithine, converted to a diamine, putrescine, through a selected decarboxylase (OrnDC). Putrescine (which can be also obtained biogenetically from arginine) is mono-N-methylated by transferase PMT and finally converted into four-N-methylaminobutanal by using diamineoxidase. Next, spontaneous cyclization-dehydration takes place, with the formation of the not unusual intermediate precursor, N-methyl-1-pyrrolinium cation, from which nicotine, cocaine and tropane alkaloids may be generated. This monocyclic precursor has further transformed into a corresponding 4-carbon side chain β-ketoacid intermediate by the action of 2 acetyl coenzyme A, (AcCoA) ester molecules. The oxobutanoic acid can cyclize to exo-carboxytropinone from which derivatives of tropine, pseudotropine and/or ecgonine are ultimately fashioned (Christen, 2000; Oksman-Caldentey, 2007).

4. Modern Pharmacology of Tropane Derivatives

Chemical compounds' relationship to pharmaceutical substances has been well-established since the turn of the 20th century and chemical synthesis has been done by improvements in properties and efficacies of natural products based upon structure-activity relationship (SAR), pharmacophore, receptor and stereoselectivity of biological action have been initiated and developed, rationale design of structure-based biological activity has thrived. Investigation of tropane alkaloids as mydriatics, spasmolytics and nearby anaesthetics have been at the frontline of newly established medicinal chemistry (Leake and Pelikan, 1976; Holzman, 1998; Sneader, 2005). Initially, new chemical analogues of prototypic compounds were biologically tested and then molecular mechanisms of biological targets (receptors and sign transduction and transmission) were developed. In contemporary pharmacology, traditional drugs and new model compounds are studied no longer best on the molecular or mobile degree, but also are presently investigated in an organismal context, the use of unique gene knockout mice as gear to observe physiological and pathological situations (Sora et al., 2001; Gomeza et al., 2002). Availability of both sorts of gear, i.e., purposefully designed compounds as molecular probes, and biological fashions for testing their pastime is equally essential for further development.

5. Pharmacology of TAs and their Role as Drug-Lead Substances

5.1. Scopolamine and Other Drugs

The TAs cocaine and scopolamine share a common tropane moiety. Nevertheless, these compounds cause very different physiological effects in humans. Cocaine manifests its effects in the synaptic cleft by inhibiting the dopamine, noradrenaline and serotonin reuptake while scopolamine acts as a competitive muscarinic receptor antagonist. The ingestion of both substances may lead to hallucinations and psychoactive effects or death (Caulfield and Birdsall, 1998; Chen et al., 2005). Hyoscyamine and scopolamine are extensively used as anticholinergic tablets. They have an effect on the central and peripheral anxious system as aggressive, non-selective muscarinic acetylcholine receptor (mAChR) antagonists that prevent the binding of the physiological neurotransmitter acetylcholine. In humans, two acetylcholine receptor sorts are regarded: Muscarinic and nicotinic receptors, which might be named after their agonists, muscarine and nicotine. Muscarine is a poison of the toadstool mushroom Amanita muscaria and acts at the mAChR of the synapses like acetylcholine, with the distinction that acetylcholinesterase does not metabolize it. The mAChRs are a subclass of the G-protein-coupled receptors (GPCRs) family, containing five subtypes (M1–M5). M1, M3 and M5 which are coupled with the stimulating Gq receptors and generate cytosolic calcium transients via the phospholipase C signalling pathway. It is assumed that those receptors are involved in brain microcirculation and mediate vasoconstriction, vasodilatation and activation of nitric oxide synthase (Henry, 1949).

Scopolamine reasons mydriatic, spasmolytic and local anaesthetics effects but famous side effects may be hallucinogenic or even lethal. The most important mode of utility for scopolamine is transdermal, Scopolamine was created as a transdermal treatment system and was used to treat motion sickness, psychotropic side effects, and shell shock during the Second World War (Dräger, 2004). Hyoscyamine and atropine have comparable modes of movement and outcomes as scopolamine. The pharmacological action of TAs is stereoselective, because of the difference in the stereoisomers concerning affinity and binding to muscarinic receptors. This outcome in specific efficiency between S-(–)- and R-(+)- isomers of hyoscyamine: The S-(–)-isomer is anticipated to be 30–300 fold more potent than the R-(+)- isomer (Husband and Worsley, 2007). The S-(−)-isomer of hyoscyamine is not solid and is racemized hastily to atropine, which is a 1:1 mixture of the two paperwork. Atropine is very strong over time and hence, it is used for medicinal packages in preference to hyoscyamine. Each atropine and scopolamine have a characteristic, dose-dependent action at the cardiovascular machine, that is clinically beneficial for resuscitation.

Cyclopentolate, which is used specifically for pediatric eye tests, and tropicamide are two other modified mydriatic sellers, which has been authorized ophthalmology considering 2005.

Tropisetron possesses a tropane skeleton however because of its mechanism of motion it belongs to the serotonin receptor antagonist. It is applied to antiemetic therapy in cases of nausea and vomiting during chemotherapy and moreover as an analgesic in fibromyalgia (Jaber-Vazdekis et al., 2006).

To limit detrimental consequences at the vital frightened machine, scopolamine has been modified by means of N-butylation and, in this form; it cannot longer bypass the blood-brain barrier. N-butylscopolamine is used to deal with stomach pain from cramping, renal colic and bladder spasms (Jung et al., 2006). Scopolamine will also be appropriate for the application in CNS sicknesses. A few research observed scopolamine to have a fast and prominent impact (Kreek et al., 2005) at the same time as others discovered no benefit from scopolamine over placebo (Leake and Pelikan, 1976) for the treatment of these situations. The contrasting findings imply that greater sizable research is needed to affirm using scopolamine for the remedy of CNS sicknesses.

5.2. Cocaine Derived Drugs

Cocaine has been used for a long term and by using many people, there is minimal information available on its usage in pharmacological treatment (Dickerson and Janda, 2005). It's far recognized that cocaine well-known shows extraordinary pharmaceutical modes of movement like nearby anaesthetic residences, CNS stimulating moves and cardiovascular outcomes. The anxious results which include euphoria, alleviation of fatigue and tedium as well as psychic stimulation. Cocaine inhibits the reuptake of dopamine, noradrenaline and serotonin, consequently growing their concentration within the synaptic cleft of the limbic gadget (Chen et al., 2005). The consumption of cocaine has a power in the brain that's detectible in an electroencephalogram (EEG). But, the consequences are inconsistent and might appear as improved or diminished signals in EEGs (Lever et al., 2005). The local anaesthetic properties of cocaine via topical utility are accomplished with the aid of blocking the ion channels in neural membranes. Cocaine is absorbed by the mucosa after application and paralysis unexpectedly happen in the peripheral ends of sensory nerves. Procaine was the first main analogue of cocaine which became particularly utilized in dentistry.

6. Important Candidates

Atropine was initially isolated from the roots of belladonna in 1831 with the help of K. Mein, a German apothecary and as a natural chemical substance by F. F. Runge in 1833 (Fodor, 1971; Sneader, 2005). Then, it took virtually half a century to research, however, this organic entity was broken up, into two components: tropine base and tropic acid.

Sooner or later, A. Ladenburg discovered that mild heating of these parts in hydrochloric acid outcomes in the restoration of the antidote structure (Sneader, 2005) (Fodor, 1971; Christen, 2000). This technique was followed by esterifying tropine with varied organic acids Ladenburg succeeded in generating a series of physiologically active compounds that he mentioned as "tropeines."

Considering one among them, the mandelic acid organic compound named homatropine (fifty-nine), acted more quickly than the antidote and had lot of less unfavourable paralytic outcomes on the ciliary muscles of the eye. Tropeines were later investigated with the help of the application of SAR however no advanced agent was found for this precise application (Ghelardini et al., 2000; Gyermek, 2002; Maksay et al., 2004).

Nowadays, the cholinergic receptor type distinguishes five metabotropic muscarinic (M1–M5) and several other teams of ionotropic nicotinic neurotransmitter receptor (nAChR) subtypes. Human muscarinic receptors (MR) are homologous proteins, 460–590 amino acids drawn-out, that belong to a taxonomic group of seven transmembrane spanning receptors which could be connected to G-protein, and consequently, their effectors are rather very well recognized (Wess, 1993; Caulfield and Birdsall, 1998; Eglen et al., 1999; Romanelli et al., 2007). The receptor proteins were cloned and various ligand affinities have been studied with the assistance of radioactive compounds (Frey et al., 1985). It's universal that MR action of tropane alkaloids is stereoselective, due to a distinction between stereoisomers in affinity and binding. The binding data processor for the neurotransmitter (and its competitive antagonists) is acidic in character. Tropane alkaloids are absorbed hurriedly from the epithelial duct. Drugs, inside the shape of an injectable, are generally administered intramuscularly. The antidote is ready to bind to mister at neuroeffector websites of muscle mass and secreter cells, peripheral ganglia and within the central nervous machine. Each antidote and hyoscine have a characteristic, dose-established movement at the vessel machine, that is clinically useful for revitalization. They inhibit secretion inside the respiration tract and additionally reduce stomachal secretion. antidote has an extended repressive impact on the channel motor activity. The movement of tropane alkaloids at the sphincter muscle of the iris and therefore the ciliary muscle of the lens, once topical applied, leads to student dilatation and to transient dysfunction of accommodation. Full healing can even take as long as seven to 12 days (Fodor, 1971; Gyermek, 1997; Christen, 2000). The antidote is likewise powerful in balancing high concentrations of ACh, which can in addition result from poisoning with organo-phosphorous pesticides or nerve gases used as chemical guns, consequently, it's applied as an Associate in nursing counter poison. It became in addition the first effective drug in the remedy of Parkinson's sickness (Gualtieri, 2000; Gyermek, 1997). Hyoscine contains an entirely similar mechanism of movement to antidote and its profile of medical use is basically overlapping. However, its action at the CNS could be a large indefinite amount bigger than mentioned, with sedation at low doses and a chance of disorientation and hallucinations discovered once at higher doses. Since hyoscine is the solely compound in the hindrance of movement illness, a percutaneous healing machine has been designed for this indication, freeing 0.5 mg of the drug over an amount of three days from a 1.5 mg reservoir. The other vital scientific package of hyoscine is premedication before anaesthesia by victimization parenteral package (i.v. infusion; s.c. or i.m. injection). Despite the actual fact that this substance contains an entirely drawn-out tale of the healthful package, reliable pharmacokinetic statistics derived from analyses of organic matrices with the assistance of liquid action with tandem bicycle mass chemical analysis detection (LC-MS/MS) device, were received simplest lately, showing sizable dependence on the indefinite quantity kind. The most drug attention in plasma happens and therefore the biological half-existence is transient. hyoscine is metabolized notably with the help of glucuronidation and by the manner in which organic compounds perform reactions (Gyermek, 1997; Christen, 2000; Renner et al., 2005; Chen et al., 2005). There's a growing body of proof that tropane alkaloids and their artificial analogues will engage properly with receptors apart from an acetylcholinergic mister.

The artificial tropeine: 1-H-indole-3-carboxylic acid organic compound of tropine, tropisetron became evolved as a selective 5-hydroxytryptamine kind 3 receptor antagonist, and its miles used clinically for remedy of surgical and chemotherapy-brought regarding vomiting (Simpson et al., 2000; Pleuvry, 2006; Husband and Worsley, 2007). Equally investigation of biological diversion of the compound realized that it may heighten glycine receptor at femtomolar concentrations, at an equivalent time as exerting repressive impact in micromolar space. Tropisetron is also a nicotinic AChR selective partial agonist. The other fulminant characteristic of this drug, is its anti-inflammatory movement, all told probability by targeting the calcineurin pathway, which is probably going to find a medical application in immunomodulation (Simpson et al., 2000). another new drug from the tropeine class is anisodamine (first extracted from Chinese language herb *Scopolia tangutica* Maxim but utilized in scientific exercise as an artificial substance), well mentioned as Associate in Nursing intermediate on propagation pathway from poison to alkaloid, that exhibited a splendid diversion as an associate in a nursing matter of pro-inflammatory cytokines, it additionally inhibits living substance aggregation and has totally different tremendous vessel moves (Fierz et al., 1974; Xiu et al., 1982; Norby and Ren, 2002).

7. Tropane-Derived Drugs in Contemporary Pharmaceutical Industry

Hyoscyamine, scopolamine and a set of their semisynthetic derivatives obtained with the help of a chemical step – N-alkylation ends up in the formation of quaternary ammonia salts (Lounasmaa and Tamminen, 1993; Gyermek, 1997; Christen, 2000). Pharmaceutical products based totally on tropane alkaloids evolved at the start of the twentieth century in Europe and progressed from the ethnopharmaceutical social group and reliable raw material supply of wild plant sources. As shortly as the meditative application of pure chemical entities began to show flavouring preparations, the necessity for nicely standardized resources of tropane alkaloids became obvious and cultivation replaced the assortment of the vegetation from wild habitats. European flora of genus Atropa and asterid dicot genus species contain on common 0.2–0.8% of overall alkaloids and share of the most loved constituent: alkaloid (hyoscine) is fairly low, that occasionally justifies business isolation based mostly all on solvent extraction. Within the beginning of the nineteenth century, phytochemical investigations distinguished that Australian vegetation, classified as Duboisia, square measure a far richer supply of tropine derivatives (Virmani et al., 1982). Currently, the non-heritability of Duboisiamyoporoides shows that closely related forms may hybridize, are cultivated on a massive scale in the Australian state, Australia, supply 10–15 plenty of contemporary leaves per area unit upon harvest, which might be created three instances twelve months. The plant is thought to comprise 2–4% general alkaloids scopolamine and constitutes the concept for delivery of the worldwide pharmaceutical enterprise (Virmani et al., 1982; Christen et al., 2007). Experimental plantations of Duboisia have already been started in India and alternative Asian nations. Biotechnological studies towards inexperienced alkaloid production are likewise ongoing however celebrated technical boundaries with bushy roots increase and process curtail the event (Mahagamasekera and Doran, 1997; Yamada and Tabata, 1997; Zarate et al., 2006). The biggest product in terms of volume, in the tropane API class, is alkaloid butylbromide and Tropisetron, launched in 1992, which could be a vital artificial antiemetic drug enforced as an associate auxiliary in cancer therapy and irradiation. It's so much nonheritable with the help of chemical esterification of tropine with indoyl chloride, among the presence of butyllithium (Simpson et al., 2000). Benztropine (antiparkinson agent used as mesylate) and deptropine (antihistaminic, administered as citrate) square measure tropine group ethers, getable from eight by victimization ballroom dancing alkylation.

Ipratropium bromide, an artificial medicine, could be a growing product with ca. 2,000 metric weight units of API artificial yearly and the corresponding value of preparations extraordinary dollar 1.7 billion! The even larger dynamic and promising drug is tiotropium bromide another artificial medicine, launched in 2002, this is utilized as a renovation treatment for patients with a prolonged, limiting pneumonic illness. Market facts indicate that tiotropium makes globally 1.2 billion dollars out of twelve metric weight units of the active substance synthesized annually (Barnes, 2000; Somand H, Remington, 2005).

Conclusion

Plant-derived tropane metabolites made a significant contribution to the history of medication – from traditional homicidal poisons to basic test compounds for physiological experiments, which is a collection of popular pharmaceutical medications. For about a century and a half, primary individual tropane alkaloids have been recognized for their pleiotropic physiological effects on both people and test animals. They are now properly identified and labelled in terms of molecular pharmacology. Numerous of these compounds are some of the high textbook examples of vital medicines derived from herbal resources. Even though their biological activities are nonselective, they may be nevertheless implemented today for many remedial disease signs. Intensive studies on tropane alkaloids chemistry and pharmacology form the foundations of important chapters of medicinal chemistry and have therefore led to new generations of structurally impressive synthetic drugs (e.g., anaesthetics and spasmolytics). Presently, industries are manufacturing tropane-based API for agricultural uses also. Both, advanced biotechnological and chemical techniques of tropane by-product synthesis is providing alternatives to plant material extraction. However, chemical synthetic processes for tropane API manufacturing are already in place for scaling up. Considering the uses of tropane alkaloids, it is very prudent to develop techniques to have more tropane-based drugs.

References

Barnes PJ (2000). The pharmacological properties of tiotropium. *Chest*, 117, 63S–66S.

Bencharit S, Morton CL, Xue Y, Potter PM, Redinbo MR (2003). Structural basis of heroin and cocaine metabolism by a promiscuous human drug-processing enzyme. *Nat Struct Biol*, 10, 349–356.

Camps P, Munoz-Torrero D: Cholinergic drugs in pharmacotherapy of Alzheimer's disease. *Mini Rev Med Chem*, 2002, 2, 11–25.

Carroll ME, Howell LL, Kuhar MJ (1999). Pharmacotherapies for treatment of cocaine abuse: preclinical aspects. *J Med Chem*, 42, 2721–2736.

Casale JF (1987). A practical synthesis of cocaine's enantiomers. *Forensic Sci Int*, 33, 275–298.

Caulfield MP, Birdsall NJ (1998). Classification of muscarinic acetylcholine receptors. *Pharmacol Rev*, 50, 279–290.

Chen H, Chen Y, Wang H, Du P, Han F, Zhang H (2005). Analysis of scopolamine and its eighteen metabolites in rat urine by liquid chromatography – tandem mass spectrometry. *Talanta,* 67, 984–991.

Chin YW, Kinghorn AD, Patil PN (2007). Evaluation of the cholinergic and adrenergic effects of two tropane alkaloids from Erythroxylumpervillei. *Phytother Res*, 21, 1002–1005.

Christen P (2000). Tropane alkaloids: old drugs used in modern medicine. In: *Studies in Natural Products Chemistry, Bioactive Natural Products*, *Part C.* Ed. Rahman A, Elsevier Science and Technology, Amsterdam, 22, 717–749.

Christen P, Bieri S, Veuthey JL (2007). Analysis of tropane alkaloids in biological matrices. In: *Modern Alkaloids. Structure, Isolation, Synthesis and Biology*. Ed. Fatorusso E, Taglialatela-Scafati O, Wiley-VCH, Weinheim, New York, 339–367.

Dickerson TJ, Janda KD (2005). Recent advances for the treatment of cocaine abuse: central nervous system immunopharmacotherapy. *AAPS J,* 7, E579–E586.

Dräger B (2004). Chemistry and biology of calystegines. *Nat Prod Rep*, 21, 211–223.

Eglen RM, Choppin A, Dillon MP, Hegde S (1999). Muscarinic receptor ligands and their therapeutic potential. *CurrOpin Chem Biol,* 3, 426–432.

Fierz G, Chidgey R, Hoffmann HMR (1974). A new tropinone synthesis. *Angew Chem Int Ed Engl,* 1974, 13, 410–411.

Fodor G (1971). The tropane alkaloids. In: *The Alkaloids; Chemistry and Physiology*, vol. XIII. Ed. Manske RHF, Academic Press, New York, 351–396.

Frey KA, Ehrenkaufer RL, Agranoff BW (1985). Quantitative *in vivo* receptor binding. Autoradiographic imaging of muscarinic cholinergic receptors. *J Neurosci*, 5, 2407–2414.

Ghelardini C, Galeotti N, Romanelli MN, Gualtieri F, Bartolini A (2000). Pharmacological characterization of the novel ACh releaser -tropanyl 2(4-bromophenyl)propionate. *CNS Drug Rev*, 6, 63–78.

Gomeza J, Yamada M, Duttaroy A, Zhang W, Makita R, Miyakawa T, Crawley J, Zhang L, Shannon H, Bymaster FP, Felder CC, Deng C, Wess J (2002). Muscarinic acetylcholine receptor knockout mice: phenotypical analysis and clinical implications.

In: *Trends in Drug Research III*. Ed. van der Goot H, Elsevier Science BV, Amsterdam, pp. 97–113.

Griffin WJ, Lin GD (2000). Chemotaxonomy and geographical distribution of tropane alkaloids. P*hytochemistry* 53, 623–637.

Gualtieri F (2000). Cholinergic receptors and neurodegenerative diseases. *Pharm Acta Helv*, 74, 85–89.

Gyermek L (1997). Tropane alkaloids. In: *Pharmacology of Antimuscarinic Agents*. Ed. Gyermek L, CRC Press, Boca Raton, pp. 47–160.

Gyermek L (2002). Structure-activity relationships among derivatives of dicarboxylic acid esters of tropine. *Pharmacol Ther*, 96, 1–21.

Henry TA (1949). The tropane alkaloids. In: *The Plant Alkaloids*. Ed. Henry TA, JA Churchill Ltd, London, pp. 69–118.

Holzman RS (1998) The legacy of Atropos, the fate who cut the thread of life. *Anesthesiology,* 89, 241–249.

Humphrey AJ, O'Hagan D (2001). Tropane alkaloid biosynthesis. A century old problem unresolved. *Nat Prod Rep*, 18, 494–502.

Husband A, Worsley A (2007). Nausea and vomiting – pharmacological management. *Hospital Pharmacist,* 14, 189–192.

Jaber-Vazdekis NE, Gutierrez-Nicolas F, Ravelo AG, Zarate R (2006). Studies on tropane alkaloid extraction by volatile organic solvents: dichloromethane. *Phytochem Anal*, 17, 107–113.

Jung D, Song J, Lee D, Kim Y, Lee Y, Hahn J (2006). Synthesis of 2-substituted 8-azabicyclo[3.2.1]octan-3-ones in aqueous NaOH solution at low concentration. *Bull Korean Chem Soc*, 27, 1493–1496.

Kreek MJ, Bart G, Lilly C, Laforge KS, Nielsen DA (2005). Pharmacogenetics and human molecular genetics of opiate and cocaine addictions and their treatments. *Pharmacol Rev,* 57, 1–26.

Leake CD, Pelikan EW (1976). An historical account of pharmacology to the 20th century (Book review). *J Clin Pharmacol,* 16, 669–671.

Lever JR, Zou MF, Parnas ML, Duval RA, Wirtz SE, Justice JB, Vaughan RA, Newman AH (2005). Radioiodinatedazide and isocyanate derivatives of cocaine for irreversible labeling of dopamine transporters: synthesis and co-valent binding studies. *Bioconjug Chem*, 16, 644–649.

Lounasmaa M, Tamminen T (1993). The tropane alkaloids. In: *The Alkaloids*. Ed. Brossi A, Academic Press, New York, 44, 1–114.

Mahagamasekera MGP, Doran PM (1997). Intergeneric coculture of genetically transformed organs for the production of scopolamine. *Phytochemistry*, 47, 17–25.

Maksay G, Nemes P, Biro T (2004). Synthesis of tropeines and allosteric modulation of ionotropic glycine receptors. *J Med Chem,* 47, 6384–6391.

Norby FL, Ren J (2002). Anisodamine inhibits cardiac contraction and intracellular Ca transients in isolated adult rat ventricular myocytes. *Eur J Pharmacol,* 439, 21–25.

Oksman-Caldentey KM (2007). Tropane and nicotine alkaloid biosynthesis – novel approach towards biotechnological production of plant derived pharmaceuticals. *Curr Pharm Biotechnol,* 8, 2003–2010.

Patterson S, O'Hagan D (2002). Biosynthetic studies on tropane alkaloid hyoscyamine in Datura stramonium; hyoscyamine is stable to *in vivo* oxidation and is not derived from littorine via a vicinal interchange process. *Phytochemistry*, 61, 323–329.

Pleuvry PJ: Physiology and pharmacology of nausea and vomiting. *AnesthInte Care Med*, 2006, 7, 473–477.

Renner UD, Oertel R, Kirch W (2005). Pharmacokinetics and pharmacodynamics in clinical use of scopolamine: *Ther Drug Monit*, 27, 655–665.

Robins R, Walton N (1993). The biosynthesis of tropane alkaloids. In: *The Alkaloids*. Ed. Brossi A, Academic Press, New York, 44, 115–187.

Romanelli MN, Gratteri P, Guandalini L, Martini E, Bonacini C, Gualtieri F (2007). Central nicotinic receptors: structure, function, ligands and therapeutic potential. *Chem Med Chem,* 2, 746–767.

Simpson K, Spencer CM, McClellan KJ (2000). Tropisetron; an update of its use in the prevention of chemotherapy induced nausea and vomiting. *Drugs*, 2000, 59, 1297–1315.

Sneader W (2005). Plant products analogues and compounds derived from them. In: *Drug Discovery; A History*. Ed. Sneader W, Wiley & Sons, Ltd, Chichester, 2005, 115–150.

Somand H, Remington TL (2005). Tiotropium: a bronchodilator for chronic obstructive pulmonary disease. *Ann Pharmacother*, 39, 1467–1475.

Sora I, Hall FS, Andrews AM, Itokawa M, Li XF, Wei HB, Wichems C, Lesch KP, Murphy DL, Uhl GR (2001). Molecular mechanism of cocaine reward: combined dopamine and serotonin transporter knockouts eliminate cocaine place preference. *Proc Natl Acad Sci USA,* 98, 5300–5305.

Virmani OP, Sharma A, Kumar A (1982). Cultivation of Duboisia as a commercial source of hyoscine and hyoscyamine – a review. *Curr Res Med Arom Plants*, 4, 47–56.

Wess J (1993). Molecular basis of muscarinic acetylcholine receptor functions. *Trends Pharmacol Sci,* 14, 308–313.

Xiu RJ, Hammerschmidt DE, Coppo PA, Jacob HS (1982). Anisodamine inhibits thromboxane synthesis, granulocyte aggregation, and platelet aggregation. A possible mechanism for its efficacy in bacteremic shock. *JAMA,* 247, 1458–1460.

Yamada Y, Tabata M (1997). Plant biotechnology of tropane alkaloids. *Plant Biotechnol,* 14, 1–10.

Zarate R, Jaber-Vazdekis NE, Medina B, Ravelo AG (2006). Tailoring tropane alkaloid accumulation in transgenic hairy roots with gene encoding hyoscyamine 6--hydroxylase. *Biotechnol Lett*, 28, 1271–1277.

Chapter 3

The Use of Alkaloids in Traditional Medicine

Rajiv Kumar[1,2,*], Shri Krishna Khandel[3] and Babita Aryal[4]

[1] University of Delhi, New Delhi, India
[2] Department of Chemistry, NIET, National Institute of Medical Science, India
[3] Clinical Diagnosis and Investigation (Rognidan), National Institute of Ayurveda, Jaipur, India
[4] Biological Chemistry Lab, Central Department of Chemistry, Tribhuvan University, Kirtipur, Kathmandu, Nepal

Abstract

Plants, herbs, and ethnobotanicals have been used for good health and disease treatment since the dawn of humans and are currently used in many parts of the world. Alkaloids are essential components of plants' defense mechanisms against pests and pathogens. Alkaloids, one element of these, are named plant secondary metabolites. They are well-known naturally occurring nitrogen-containing bioactive compounds. Modern research is being conducted on alkaloids to find novel therapeutic approaches. Many biological processes are known to be aided by alkaloids, some of which can also transform into active metabolites. Numerous natural sources and the synthesis of different alkaloids with significant therapeutic effects are the sources of the hundreds of medicines that are successfully employed in the context of health and the treatment of various disorders. In this regard, this chapter discusses alkaloids used as medicines with scientific and historical evidence and covers plant-derived alkaloids and their therapeutic activities, including acetylcholinesterase (AChE) inhibitory activity, anti-inflammatory activity, and antioxidant activity. Moreover, the use of alkaloids in

---

* Corresponding Author's Email: chemistry_rajiv@hotmail.com.

In: The Essential Guide to Alkaloids
Editor: Deepak Kumar Semwal
ISBN: 979-8-88697-456-0

treating neurodegenerative and inflammatory diseases has been included. Further, the authors cover therapeutic applications of plant-derived alkaloids, explain mediators of neurodegenerative disorders, and inflammatory mediators, explain the therapeutic potential of alkaloids against human coronaviruses, discuss the usage of alkaloids in current medicine and describe plant-derived alkaloids (tetrahydropalmatine, aloperine, tetrandrine, sinomenine, oxymatrine, galantamine, berberine, and harmine); write sections covering mechanisms and methods of treatments for various diseases, elucidate the active use of the alkaloid boldine in traditional medicine as a natural antioxidant; and in the end, the odd brominated alkaloids derived from marine sources are intensely conferred. The ability of alkaloids to form hydrogen bonds with enzymes, receptors, and proteins is also examined. The role of alkaloids in curing numerous illnesses, including Huntington's disease, multiple sclerosis, Alzheimer's disease, Parkinson's disease, anxiety, depression, neurotoxicity, cerebral ischemia, cognitive decline, dementia, drug addiction, autoimmune encephalomyelitis, brain cancer, psychiatric disorders, and epilepsy have also conversed.

Keywords: herbal medications, novel drug delivery system, ayurvedic formulations, natural products

1. Introduction

A group of naturally occurring organic nitrogen-containing substances known as alkaloids is regularly observed in the plant kingdom. In higher plants from the African genera Apocyanaceae, Ranunculaceae, Rutaceae, Papaveraceae, and Solanaceae, alkaloids are extensively dispersed (Patwardhan, 2005). Additionally, reports of them in marine species, microbes, lower plants, and insects have been made. Alkaloids are used in ancient and modern medicines in amounts ranging from 25 to 75 percent, which highlights their enormous pharmacological importance and significant therapeutic potential. Alkaloids are tiny chemical compounds that are secondary metabolites of plants and often include nitrogen in a ring; they make up roughly 20% of plant species (Krikorian, 1998). Although this is frequently the case, it is not always the case that the ring is aromatic (a ring in which the electrons can be shared among the complete ring structure) (Harvey, 2008). Alkaloids often have neutral nitrogen and can take hydrogen because of this. At least one nitrogen will always be present in the molecule and all alkaloid structures feature rings. The nitrogen will become neutral after three bonds, and it will become positively charged after the fourth bond. Alkaloids have nitrogen atoms with a neutral configuration making them a good base by enabling them to accept extra hydrogen. Therefore, N can have three or four bonds. Alkaloid structures can differ significantly from one another, aside from a ring and nitrogen. For thousands of years, people have used plant alkaloids (either as poison or medicine). The formal definition of an alkaloid nowadays, when the pure structures of alkaloids can be determined, is a cyclic molecule containing nitrogen, which is present in living organisms.

Alkaloids can be discovered in food and beverages since they are present in a wide variety of plants. Thebaine, heroin, noscapine, and papaverine are some other alkaloids with comparable structural features. Caffeine found in tea, coffee, and chocolate is one of the food alkaloids. Alkaloids (hirsuteine, hirsutine, atropine, berberine, isorhynchophylline, jatrorrhizine, isocorynoxeine, and palmatine) as well as the harmful effects of solanine, nicotine, anabasine, sanguinarine, and aristolochic acid I) often have physiological effects on people (Lisko et al., 2017). These physiological impacts include both mild stimulants like coffee and toxins like coniine (the poison in hemlock). Thebaine, heroin, noscapine, and papaverine are some other alkaloids with comparable structural features (Singh et al., 1989). Other alkaloids, like morphine and oxycodone, exhibit striking structural similarities with only minor changes (Figure 1) (Aryal et al., 2022). Small levels of poisonous alkaloids, such as solanine in potatoes and tomatoes, are present in certain widely consumed foods.

The greener part of the potato contains more solanine. However, even the majority of green potatoes do not have enough solanine for the body to have any effects. However, if enough was consumed, it might result in coma and death, as well as nausea, vomiting, and diarrhoea. Alkaloids' frequently low cytotoxicity and diversity in converting into stable salt have led to a variety of medications that are simple to administer in the body without the negative effects brought on by ingesting organic and inorganic salt with poor tolerance. Alkaloids are also used to describe several synthetic substances with related structural features (Bharadwaj et al., 2018). While some alkaloids are free bases, others combine with organic acids like acetic and oxalic acid to create salts. Amino acids, which supply the necessary nitrogen for the alkaloid structure, are the beginning point for the synthesis of alkaloids in plants. Alkaloids play a key role in plants' defensive mechanisms against diseases and herbivores. The hundreds of medicines that are successfully used in the context of health and the treatment of various diseases have their roots in various natural sources as well as the synthesis of several alkaloids with significant therapeutic action.

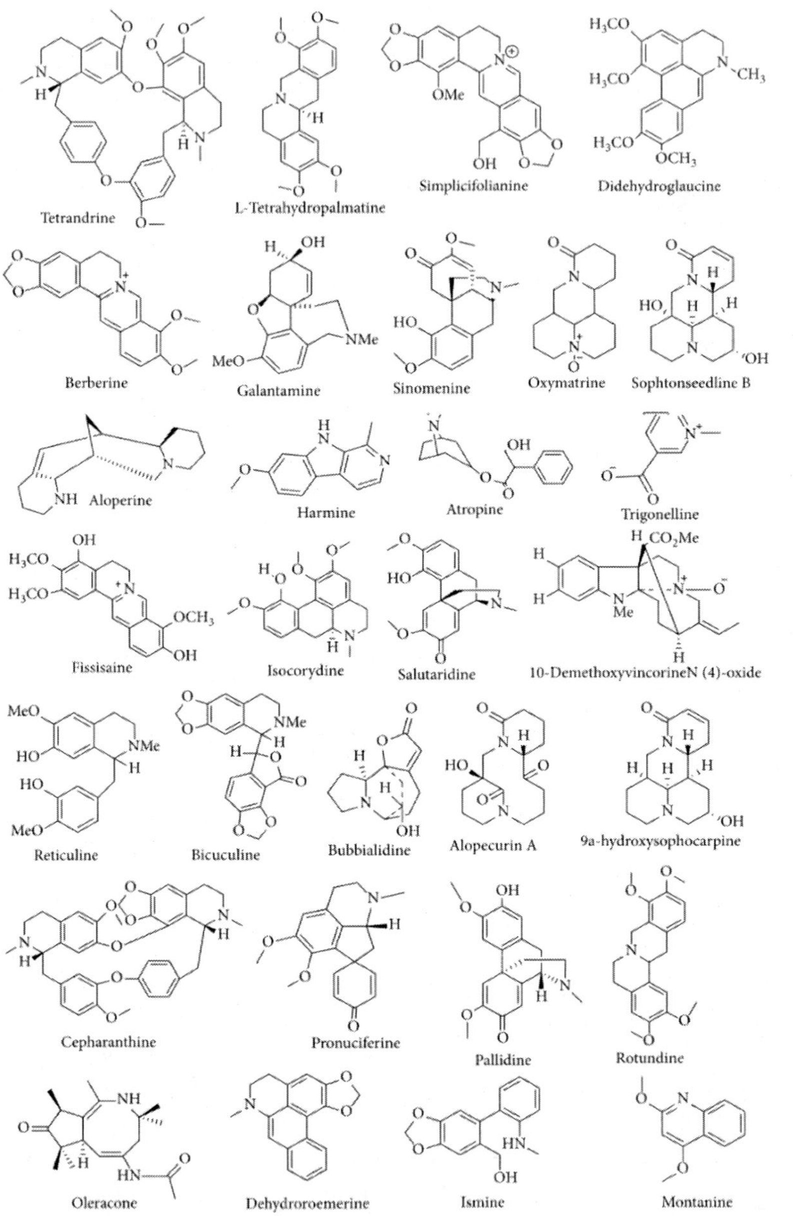

Figure 1. Molecular structures of alkaloids represent how all alkaloids have a ring and at least one nitrogen atom.

Numerous alkaloids are effective drugs that can be used to treat a wide range of illnesses, including malaria, heart failure, diabetes, cancer, etc. Similar to this, the root cause of disorders related to blood clotting is platelet aggregation which occurs for reasons other than homeostasis (Schmeller and Wink, 1998). Alkaloids are classified as local anaesthetics, stimulants, analgesics, antibacterial, psychedelics, stimulants and anticancer medications, as well as antiarrhythmic, antihypertensive, cholinomimetic, spasmolysis, antiasthma, vasodilator, antimalarial, and other therapeutic alkaloids in medicine (Sneader, 2006). The protective function, detrimental health impacts, and the therapeutic qualities of atropine, hirsutine, isorhynchophylline, berberine, hirsuteine, isocorynoxeine, palmatine, and jatrorrhizine as well as the harmful effects of anabasine, aristolochic acid I, sanguinarine, solanine, and nicotine. Some plant alkaloids, such as solanine in solanum, are found in a glycosidic form (Wink, 1998). Particularly, the decarboxylation of substances involves the alkaloid biosynthesis route. The antiplatelet and anticoagulant alkaloids found in medicinal plants are abundant. Salicin, a substance produced from the willow plant and frequently used in pain management, is the source of the widely used antiplatelet drug aspirin.

Plant secondary metabolites are known as alkaloids (Aniszewski, 2007). They are well-known naturally occurring bioactive chemicals that include nitrogen. Alkaloids are the subject of cutting-edge research to uncover new medicinal strategies. Alkaloids are known to contribute to a variety of biological processes, and some of them can also change into active metabolites (Aniszewski, 2007). We have concentrated on commercially available and unproven alkaloids in this discussion. We have included the origins and biological effects of documented alkaloids from earlier decades. This comprehensive review of the chapters, which covers the clinical treatment of diseases of various types as well as the future development of new vaccines, will help us know and understand the significance and magnitude of the alkaloids and their role in our lives. These chapters include a variety of nitrogenized compounds, such as marine and terrestrial alkaloids. A class of organic compounds that have been employed by civilizations throughout history are addressed here and are still used today for a variety of medical and therapeutic purposes. These compounds are frequently mentioned in bibliographic references. As a result, alkaloids have been isolated from both marine and terrestrial sources, and humans have developed the ability to analyze the chemical makeup of numerous derivatives, both simple and complex, as well as to observe how each component affects living things biologically.

The basic characteristics of alkaloids are no longer necessary for an alkaloid, and at least four different classes of nitrogenous compounds are permitted by the chemistry of nitrogen atoms (Davis and Roughley, 2017). The majority of alkaloids are challenging to create in a lab. Instead, as has been the case for thousands of years, alkaloids are typically extracted from plants. Although modern techniques can produce far more thoroughly isolated and pure alkaloids than those employed in the past, many alkaloids are still quite challenging to do so. One of the few alkaloids that are frequently synthesized is caffeine, and most of the caffeine in soda, other energy beverages, and foods comes from this process. This study discusses several alkaloids with an antiplatelet activity that have been extracted from plant sources, along with potential mechanisms and candidates for additional in-depth research in the search for antiplatelet medicines (Hall and Mazer, 2011; Debnath et al., 2018). About 80% of the populace in underdeveloped nations is thought to primarily rely on plant-based traditional healing to meet their basic healthcare needs.

Traditional remedies are typically made from various medicinal plant components rather than the entire plant (Yang and Stöckigt, 2010). Despite opposition, traditional medicine has had a significant renaissance for a long time (Patnala and Kanfer, 2010). Traditional healers play a crucial role in supplying leads for the identification of pharmacologically active plant-derived chemicals, in addition to providing immediate healthcare to the rural population. This chapter provides a full grasp of alkaloids as antiplatelet agents with a potential mode of action (Figure 2). The antiplatelet activity of alkaloids and their therapeutic use as effective antiplatelet medicines will also be covered here, along with a description of structural relationship activity and potential lead compounds for further drug discovery (Ain et al., 2016).

Each section of this chapter address and explore the issues they create for the future, particularly in the areas of feeding and health in general (Qiu et al., 2014). In this regard, this contribution includes alkaloids used as medicines: scientific and historical evidence, plant-derived alkaloids and their therapeutic activities, acetylcholinesterase (ache) inhibitory activity, anti-inflammatory activity, antioxidant activity, neurodegenerative and inflammatory diseases: therapeutic applications of plant-derived alkaloids, mediators of neurodegenerative disorders, inflammatory mediators, alkaloids: therapeutic potential against human coronaviruses, utilization of alkaloids in current medicine, and promising plant-derived alkaloids (tetrahydropalmatine, aloperine, tetrandrine, sinomenine, oxymatrine, galantamine, berberine, and harmine); sections on mechanisms and treatments for various diseases, where various alkaloid types govern significant and specific forms; the active use of the alkaloid boldine in traditional medicine as a natural antioxidant; the odd brominated alkaloids from marine sources; alkaloids of vegetable origin as coming from the Amaryllidaceae; among other notable examples; alkaloids derived from the Erythrina, including their synthesis and medicinal uses; some of their derivatives' technological methods; ovarian cancer treatment with Trabectedin; discussion of a tiny set of alkaloids termed oxoisoaporphines as a major medical tool in the treatment of mental disorders including depression; and finally, a thorough examination of the Daphniphyllum alkaloids are all interesting contributions.

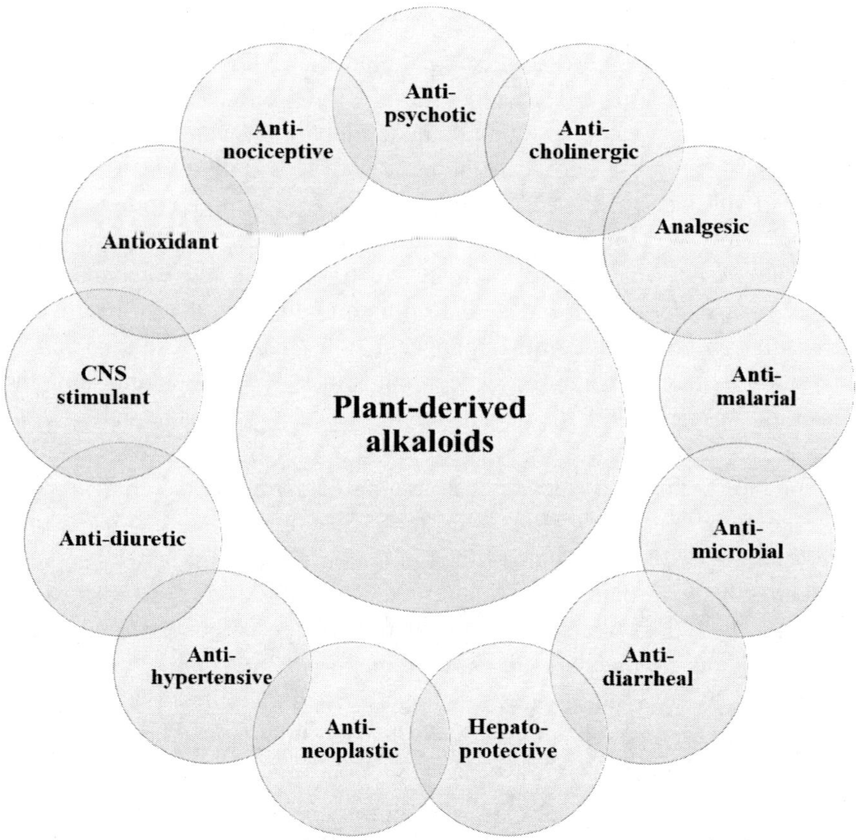

Figure 2. Biological activities of different plant-derived alkaloids.

2. Alkaloids as Medicines: Scientific and Historical Evidence

As old as humanity itself, medicinal plants have been used to treat illnesses. Written accounts, historical monuments, and even the original plant medicines provide compelling evidence of man and his long-standing quest for natural cures. Man's long-standing battles with disease prompted him to hunt for pharmaceuticals in the barks, seeds, fruit bodies, and other sections of plants, which led to the understanding of employing therapeutic plants. On a Sumerian clay slab (Nagpur), which is considered to be about 5000 years old, was found the earliest known written account of the utilization of medicinal herbs for medication manufacture (Tembo et al., 2021). There were 12 recipes for making drugs that mentioned more than 250 different plants, some of which had alkaloids. In the Chinese book "Pen T'Sao" on roots and grasses, which was penned by Emperor Shen Nung around 2500 BC, there are references to 365 drugs (dried medicinal plants and their parts) (Hou, 1977). These include *Rhei rhizoma*, Podophyllum (yellow gentian), camphor, folium, ginseng, jimson weed, ephedra, and cinnamon bark. The ancient texts known as the Vedas reference herbal medicine, which is widely used in India. The Ebers Papyrus, a collection of 800 prescriptions, was written around 1550 BC. Approximately 700 plants and their different species and medications are listed, including pomegranates, juniper, castor oil plants, aloe, senna, garlic, onions, figs, willows, coriander, and common centaury, that are utilized as therapeutic agents (Hallmann-Mikołajczak, 2004). Homer's The Iliad and The Odysseys, written circa 800 BC, contain allusions to 63 plants and their concerned species from the Assyrian, Mycenaean, Minoan, and Egyptian cultures.

Some of them received names in honour of mythological characters that were mentioned in epics; for occurrence, Elecampane (*Inula helenium* L. Asteraceae) was given the name Elena in honor of the goddess who served as the centre of the Trojan War (Cantrell et al., 1999). In 500 BC, Herodotus referred to the castor oil plant, Orpheus to the scented hellebore and garlic, and Pythagoras to the mustard, sea onion (*Scilla maritima*), and cabbage (Kurhekar, 2020). Hippocrates' writings, which date from 459-370 BC, include 300 therapeutic plants in order of their physiological actions (Patwardhan, 2005). Theophrast (371-287 BC) established botanical science and named his works "De Causis Plantarium" Plant Etiology and "De Historia Plantarium" Plant History (Petrovska, 2012). In his book "De re medica," the renowned physician and author Celsus (25 BC-50 AD) referred to almost 250 medicinal plants, including aloe, cinnamon, pepper, henbane, the star gentian, poppy, flax, cardamom, false hellebore, etc.

Natural cures for ailments have been sought after since the dawn of mankind. Similar to animals, people used medicinal plants initially out of pure instinct (Nigam, 2020). Because there was not enough information at the time about the origins of the illnesses or the exact plants that may be used as medicine, everything was based on experience. Several plant-based medicines have been used for thousands of years and were known to ancient societies that are included in modern pharmacology today (Viljoen, 2013). Modern science has acknowledged its functional impact. Due to their understanding of the advancement of concepts relating to the use of medicinal plants as well as the growth of consciousness, pharmacists and doctors are better able to respond to problems that have arisen with the proliferation of professional services in the facilitation of man's life. The usage of medicinal plants increasingly renounced the empiric framework and was instead based on explicative facts as the rationale for employing specific medicinal plants to treat specific conditions emerged throughout time. Before the development of iatrochemistry in the sixteenth century, treatment and prevention were derived from plants. However, given the declining effectiveness of synthetic drugs and the expanding list of contraindications to their usage, the use of natural pharmaceuticals is once again a hot topic.

Historically, the market performance of medicinal alkaloids has been connected with a species' abundance as described by the GBIF dataset, the largest database currently accessible specifying the distribution of individual species (Amirkia and Heinrich, 2014). Nevertheless, alkaloids are not usually utilized as lead compounds for the promotion and approval of innovative medicines. This analysis is based on 117,909,945 records (or 27.8% of all entries for all organisms) from the kingdom Plantae. Since 2014, 204,963,126 more records, or 3.82 times as many, have been added to GBIF.

It will be beneficial on several levels to combine metabolomics technologies with techniques for finding natural compounds. In the first place, by increasing the number of identifications in our metabolomics data, we might be able to provide novel structures that can be tested for bioactivity for any ailment that is being studied. With multi-parallel analysis using metabolomics technology, chemical characterization methods of numerous varied species from natural resources will be improved and increased throughput. Second, natural product chemists have amassed a lifetime's worth of compound libraries of both active and inert pure compounds that can now be investigated to generate mass spectrum and NMR spectral libraries, increasing biological interpretations of metabolomics data (Emwas et al., 2019). Analytical instrumentation improvements and sophisticated hyphenation of separation techniques with high sensitive detectors have enabled greater detection of small molecule compounds measurable in biological systems. These advancements are helping to advance the discovery of natural product chemistry and identify potential novel drug candidates that will help maintain health and fight disease (i.e., primary and secondary metabolites).

3. Plant-Derived Alkaloids and Their Therapeutic Activities

Plant-derived alkaloids are expected to be beneficial therapeutic agents for disease treatment or drug administration. These nitrogenous chemicals known as alkaloids have a variety of biological effects. Alkaloids, which are largely found in the plant kingdom, are simple nitrogen-containing chemical compounds that are produced from amino acids (Cronin, 1990). These therapeutic target specific tissues through special endocytosis pathways, skewing them towards diseased cells while being remarkably biocompatible and only mildly toxic to healthy cells (Figure 3). As a result, using these plant-based alkaloids may broaden the scope of available drug therapy while reducing adverse effects. Alkaloids are poisonous substances that plants make and build up as a result of abiotic stresses, biotic pressures, and environmental changes. Alkaloids can form stable salts with better pharmacokinetics than non-basic drugs in this situation.

Experimental paradigms in vitro and in vivo settings have been employed to describe the proinflammatory factor-inhibiting properties of marine alkaloids. The use of various plant-derived alkaloids as medications, CNS stimulants, psychedelic substances, and poisons has long piqued attention and had an impact on human history (Rajput et al., 2021). In cellular and animal models, plant alkaloids have shown a variety of medicinal and pharmacological effects, including neuroprotective, antitussive, anticancer, anti-inflammatory, cardioprotective, vasorelaxant, antifungal, antiparasitic, antibacterial, and antiprotozoal properties. The biological effects of alkaloids, including their anticancer, antimicrobial, antioxidant, AChE inhibitory, antibacterial, anti-inflammatory, antidiabetic, and antimalarial activities, have been studied and reported in both conventional and current medical schemes (Thawabteh et al., 2019). The diagrammatic graph below illustrates the biological claims mentioned in the literature for alkaloids.

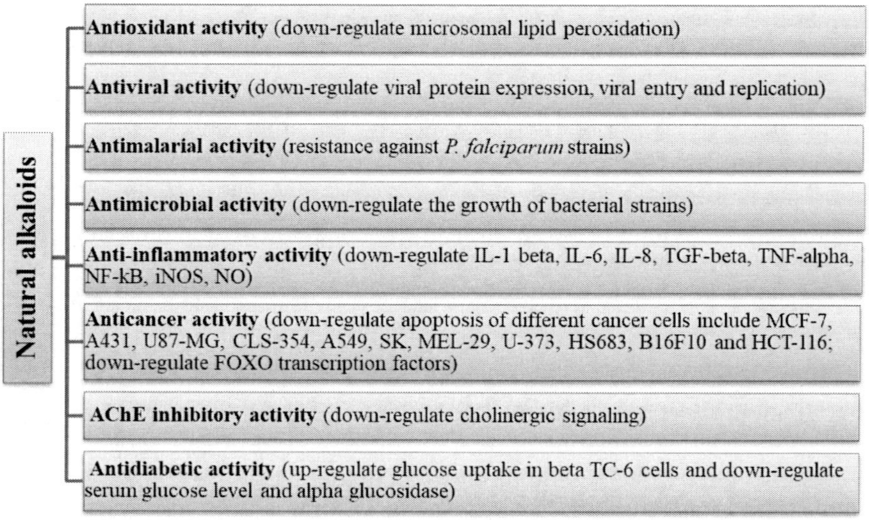

Figure 3. Important biological activities of alkaloids with their mechanisms. Abbr. IL - interleukin, TGF - transforming growth factor, TNF - tumour necrosis factor, NF-kB - nuclear factor kappa, iNOS - inducible nitric oxide synthase, NO - nitric oxide, MCF-7 - Michigan Cancer Foundation-7, U87-MG - Uppsala-87 malignant glioma.

Alkaloids are also examined for their propensity to establish hydrogen bonds with enzymes, receptors, and proteins due to the presence of protons receiving N-atoms and one or more protons giving amine H-atoms. Alkaloids have been shown to have neuroprotective effects against diseases like Huntington's disease, multiple sclerosis, Alzheimer's disease, Parkinson's disease, anxiety, depression, neurotoxicity, cerebral ischemia, cognitive decline, dementia, drug addiction, autoimmune encephalomyelitis, brain cancer, psychiatric disorders, epilepsy, and others. Alkaloids have the potential to act as disease-modifying agents because of their knowledge of modulating a variety of biochemical and molecular markers regulating several signal transduction pathways, which is related to the complexity of neurological illnesses. The majority of alkaloids originating from plants have been found to have antibacterial, insecticidal, antimetastatic, antiproliferative, and antiviral properties (Matsuura and Fett-Neto, 2015). In this chapter, the authors discuss the possible therapeutic applications of plant-derived alkaloids that have been extracted from various plant species. The authors also highlight the key properties and biological consequences of these compounds.

3.1. Acetylcholinesterase Inhibitory Activity

A promising approach for treating numerous health complications is the inhibition of acetylcholinesterase (AChE), the important enzyme in the breakdown of acetylcholine (Zhan et al., 2020). The plethora of plants in nature undoubtedly serves as a possible source of AChE inhibitors. In this communication, numerous phytoconstituents and potential plant species such as AChE inhibitors are presented. AChE was recognized as a key player in preventing NDDs from impairing cholinergic transmission. Nitidine and avicine, two alkaloids, isolated from *Zanthoxylum rigidum* have been showed AChE inhibitory activity (Stavrakov et al., 2020). Moreover, four pyrrolizidine alkaloids, including heliosupine, 3′-O-acetylheliospine-N-oxide, and 7-O-angeloylechinatine-N-oxide isolated from *Solenanthus lanatus* displayed good AChE inhibitory activity. Because vobasenal-type monoterpene indole alkaloids have an N-methyl group and interact with Trp133 and Trp86 moieties be present at hydrophobic appendages, rauvomitorine III alkaloids among those extracted from Rauvolfia vomitoria leaves exhibited anti-AChE activity. Likewise, the bark of Holarrhena pubescens included the alkaloids mokluangin A-C and antidysentericin, which both had potent AChE inhibitory action.

According to chemical fingerprints, the most potent AChE inhibitors include alkaloids like galantamine, lycoramine, caranine, and N-demethylgalanthamine found in the leaves, bulbs, and roots of Amaryllidaceae plants like Crinum, Zephyranthes, and Habranthus (Cheenpracha et al., 2016). Through a combinatorial library, Galantamine was discovered dual-site binding AChE inhibitor. According to these results, plant-derived alkaloids exhibit a strong anti-AChE inhibitory effect (Figure 4). The major enzyme in the breakdown of acetylcholine, AChE, is thought to be a viable therapeutic approach for the management of neurological illnesses such as Alzheimer's disease, myasthenia gravis, senile dementia, and ataxia (Shawky et al., 2021). The plethora of plants in nature is undoubtedly a potential source of AChE inhibitors (Arya et al., 2021). The goal of the authors is to present a thorough literature review of plants that have been evaluated for their ability to inhibit AChE.

In this section, numerous phytoconstituents and potential plant species such as AChE inhibitors are presented. Strong AChE inhibitory action was reported for the extracts of *T. polium, M. longifolia, Thymus vulgaris, Mentha x piperita, T. chamaedrys, Satureja montana, T. montanum, Teucrium arduini, and Salvia officinalis*. According to spectrophotometric analysis, the total amount of hydroxycinnamic compounds in the extracts varied significantly (Vladimir-Knezevic et al., 2014). With the exception of *Teucrium* species, rosmarinic acid was discovered to be the main component in the majority of the medicinal plants under investigation and to have a significant impact on their AChE inhibitory and antioxidant capabilities.

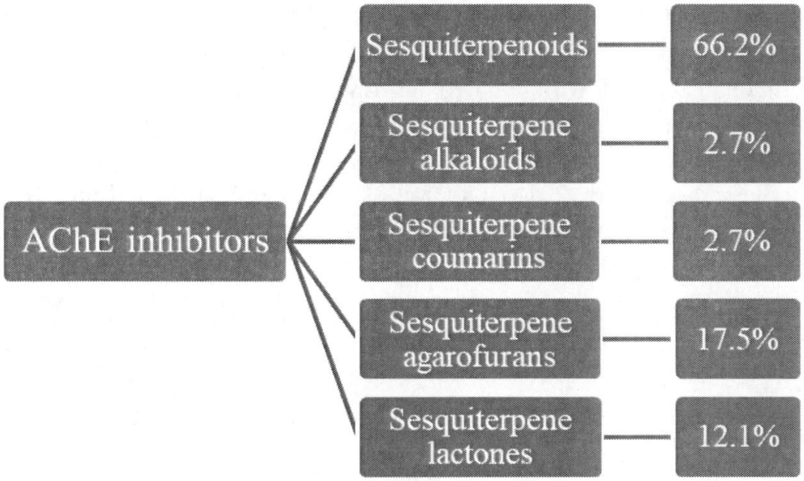

Figure 4. Percentage of sesquiterpenes having the ability to inhibit AChE in different categories.

3.2. Anti-Inflammatory Activity

The traditional folkloric use of botanicals containing alkaloids that is described in traditional literature and several pharmacopoeias has been supported by bioactivity-guided isolation of numerous neuroprotective alkaloids. The structure-activity connection in alkaloids delivered alone, in combination with other natural or synthetic medications, or as ingredient(s) in polyherbal formulations may shed light on the underlying method of action of plant alkaloids because synergism exists among phytochemicals (Abat et al., 2017). However, there haven't been many investigations on the structure-activity link regarding the neuroprotection provided by alkaloids (Wansi et al., 2013). The genesis of neuroprotective plant alkaloids and their underlying mechanism as therapeutic agents shown in vitro, in vivo, ex vivo, or human clinical research are described in the current work. Alkaloids will thus be synthetically optimized as disease-modifying agents as a result of the massive structural diversity of conditions that by nature and lead-identifying approaches combined with contemporary integrated approaches like metabolic engineering, robotic separation, genetic sequencing, structure-activity studies, and synthetic biology (Satake et al., 2013). For instance, nitrogenous chemicals called alkaloids, which are produced from amino acids, have a variety of biological activities.

In numerous investigations, compounds with distinct chemical properties from those found in plant-derived alkaloids are active in a variety of experimental settings, with detailed descriptions of their structures, activities, and mechanisms of action. Proteins and peptides, carotenoids, phenols and polyphenols, and polysaccharides all have well-known anti-inflammatory properties (Yatoo et al., 2018). Other classes of chemicals in addition to these substances have sparked interest in this field of bio-prospecting. Plant alkaloids are a significant family of compounds having anti-inflammatory activity and can inhibit several pro-inflammatory factors, including cytokines, lipid mediators, histamine, and inflammatory response enzymes. Alkaloids, which are mostly weak bases, are better therapeutic candidates than compounds with a non-ionization profile due to their ionization profile. Additionally, basic groups including amines, guanidine, amides, and amidine, can procedure salts in biological conditions. To create molecules with a low hydrophobicity, these groups are added.

A medication or chemical reduces inflammation (redness, swelling, and pain) in the body. Anti-inflammatory medications operate by limiting the body's production of specific chemicals that are responsible for inflammation. Alkaloids are employed to treat a variety of diseases. Some anti-inflammatory medications are being studied for use in the treatment and prevention of diseases. Infectious microorganisms like bacteria, viruses, or fungi that invade the body, settle in particular tissues or circulate in the blood, produce inflammation. Additionally, tissue injury, degeneration, cell death, and cancer, ischemia can all result in inflammation. In most cases, the development of inflammation is attributed to both innate and adaptive immune responses (Zhao et al., 2018). The most crucial defense against invasive bacteria and cancer cells is the innate immune system, which is composed of macrophages, mast cells, and dendritic cells. In the adaptive immune system, more specialized cells, such as B and T cells, produce specific receptors and antibodies that are used to fight encroaching pathogens and cancer cells.

Platelets are little nuclear cells with an abnormal shape (thrombocytes). When platelets are stimulated, they take on a distinct form from their resting discoid shape. Platelets rush to the site of an injury when blood vessels are damaged or harmed and form aggregates to stop the bleeding. Their attachment to the exposed active thrombin receptor enables this. Blood platelets are a component of normal homeostasis. The usual homeostatic system limits blood loss by regulating interactions between blood platelets, plasma proteins, and components of vessel walls. Platelets have a critical role in the development of cardiovascular diseases. Arterial thrombosis, an acute complication that manifests on chronic atherosclerotic lesions, causes heart attacks and strokes (Badimon and Vilahur, 2014). These chronic inflammatory processes are the primary pathophysiological mechanism by which lipid buildup provides substrates for the development of occlusive thrombus (Dall'Acqua, 2013; Furman et al., 2019). Since several drugs used to treat inflammation may unintentionally reduce normal platelet function, there is always a need for alternatives to treat such conditions.

Numerous inflammatory mediators are produced and released during different types of inflammatory reactions. The two main categories of inflammatory molecules are pro- and anti-inflammatory mediators. However, some mediators, including interleukin (IL)-12, have both pro- and anti-inflammatory properties (Scheller et al., 2011). Several inflammatory mediators and cellular pathways have been thoroughly investigated in connection with human pathological conditions, including cytokines (e.g., interferons, eicosanoids (e.g., prostaglandins and leukotrienes), interleukins, and tumour necrosis factor), chemokines (e.g., monocyte chemoattractant protein 1), and the potent inflammation-modulating transcription factor nuclear factor B (Turner et al., 2014). The authors of this chapter have compiled the most recent information on the anti-inflammatory properties of alkaloids from marine species (Ngo et al., 2017). For example, canthin-6-one alkaloids from the stem bark of Ailanthus altissima exhibit an anti-inflammatory impact by reducing Akt phosphorylation and NF-B transcriptional activations.

Berberine, a plant-derived alkaloid, has shown promise in the treatment of Acne Vulgaris by reducing pro-inflammatory cytokines such IL-1, TNF-, IL-6, and IL-8. Similar to this, the ethanol extract of 1-carbomethoxy-carboline alkaloids extracted from *Portulaca oleracea* demonstrated the strongest anti-inflammatory activity because it inhibited NF-B and the MAPK pathways, which reduced the generation of pro-inflammatory mediators like inducible IL-6, IL-1, nitric oxide synthase, TNF-, and induced IL-6 (Lou et al., 2011). A plant-derived indole alkaloid called rhynchophylline (Rhy) was discovered in the *Uncaria* species. Both the plant and the alkaloid have several defensive qualities, including sedative, anti-inflammatory, neuroprotective, and anti-hypertensive actions. Several studies back up the significance of the plant's anti-inflammatory effect as the underlying mechanism for the majority of the alkaloid's pharmacological benefits. Rhy is good at defending the cardiovascular and central nervous systems. Changes in living choices are the primary cause of cerebro-cardiovascular disease. Numerous earlier studies have shown the importance of Rhy in controlling calcium and potassium channels, hence defending the brain from neurodegenerative disorders and their associated side effects (Hussain et al., 2018). Additionally, Rhy exhibits anticoagulant and antiplatelet aggregation properties.

3.3. Antioxidant Activity

Numerous medicinal plants have had their antioxidant capacities studied (Krishnaiah et al., 2011). Exogenous (dietary, i.e., plant-based) antioxidants have long been found in plants. Two-thirds of all plant species on earth are thought to have medicinal value, and almost all of them have excellent antioxidant potential (Aryal et al., 2021b). The discovery and subsequent separation of ascorbic acid from plants originally sparked interest in the exogenous plant antioxidants. Since then, increasing oxidative stress has drawn a lot of attention because it is a major contributing element in the onset and progression of many serious diseases, including cardiovascular and neurological disorders (Tafrihi et al., 2021). Additionally, it has been discovered that strengthening the body's natural antioxidant defense or taking supplements of exogenous antioxidants is a potential way to combat the negative consequences of oxidative stress. For the evaluation of the antioxidant activity of plant samples, the majority of these in vitro tests revealed strong antioxidant activity in plant samples. This is most likely because of their natural capacity to produce secondary metabolites like phenolic compounds as well as non-enzymatic antioxidants like ascorbic acid and glutathione.

Researchers have found antioxidants like phenolics, proanthocyanidins, flavonoids, and tannins in studies on herbal plants, vegetables, and fruits. The anti-inflammatory properties of medicinal plants may contribute to the disease defense they provide. Consuming natural antioxidants has been shown to have a negative relationship with morbidity and mortality from degenerative diseases. A severe health issue is liver disease. Free radicals are known to harm cells by covalent binding and lipid peroxidation, which results in tissue injury (Lobo et al., 2010). Natural antioxidants have drawn particular attention due to their capacity to scavenge free radicals. Hepatic damage has been linked to the use of medicinal plants with high antioxidant content as a treatment strategy. Antioxidants, when present in lower quantities than the substrate, significantly slow down or stop the oxidation of oxidizable substrates. Antioxidants can be consumed or produced in the body (e.g., GSH, SOD, etc.). Natural antioxidants are highly good at stopping the harmful effects of oxidative stress, whether they are in the form of raw extracts or their chemical components.

Free radical reactions are known to play a significant part in the pathophysiology of many acute and chronic human diseases, including diabetes, immunosuppression, atherosclerosis, ageing, and neurodegeneration (Feng et al., 2018). The body's natural antioxidant capacity and ROS levels were out of balance, which led to the requirement of dietary and/or pharmacological supplementation, especially during a disease assault. It is commonly acknowledged that medications made from plant products are safer than their synthetic counterparts, even though the toxicity profile of the majority of medicinal plants has not been properly examined. Reactive oxygen species (ROS) and other oxidants have been linked to several ailments and diseases, according to a large body of research (Jiang et al., 2018). The findings have drawn scientists' attention to the value of antioxidants in the prevention and treatment of diseases as well as in maintaining human health. The human body has an innate antioxidative mechanism, and many biological processes like the defense against mutagenesis, ageing, and carcinogenesis are based on this characteristic. Free radicals are generally stabilized or inactivated by antioxidants before they may damage targets in biological cells.

The outcomes of this research are then immediately extrapolated to the medicinal value of the phytochemicals. Nevertheless, phytochemicals are being evaluated for their in vitro antioxidant activity (Obaid Aldulaimi et al., 2019). The relevance of plants as exogenous sources of antioxidants and the effectiveness of their medicinal properties may be seriously questioned in light of this misconduct. As a result, we covered briefly in the current paper the physiology and redox biology of both plants and people (Müller-Schüssele et al., 2021). Discussion is also had regarding the uses and restrictions of assays used to detect antioxidant activity. The knowledge presented here will allow accurate interpretation of the results of research assessing the antioxidant potential of plants using both in vitro and in vivo experiments. The utilization of naturally occurring antioxidants in food, pharmaceuticals, and cosmetics has recently seen a significant uptick in interest due to their versatility in terms of their range and intensity of activity as well as their vast potential for redressing the imbalance. Additionally, in vitro tests have revealed that several phytochemicals exhibit antioxidant activity (Vera-Reyes et al., 2015; Ashraf, 2020). However, due to their interference with physiopharmacological processes like absorption, distribution, metabolism, storage, and excretion, only a small number of compounds have been demonstrated to be therapeutically helpful in vivo settings.

Multiple diseases' onset and progression have been linked to oxidative stress as their primary cause. Exogenous antioxidant supplementation or strengthening the body's endogenous antioxidant defences are promising ways to counteract the negative effects of oxidative damage brought on by reactive oxygen species (Sharma et al., 2012). Plants have a natural potential to biosynthesize a wide range of non-enzymatic antioxidants capable of attenuating ROS-caused oxidative damage. Plants have been screened for their antioxidant potential using a variety of in vitro techniques, and in the majority of these assays, powerful antioxidant activity was found. Plant antioxidants must first undergo several physiopharmacological procedures to demonstrate their in vivo therapeutic efficacy (Sharma et al., 2012). As a result, the results of antioxidant potential evaluation tests conducted in vitro and in vivo are not always consistent. However, without conducting enough in vivo research, the outcomes of in vitro experiments have been irrelevantly extended to the therapeutic application of plant antioxidants.

To enhance our understanding of plant antioxidants as therapeutic agents, the authors have therefore briefly examined the physiology and redox biology of both plants and humans. To determine the exact course to be taken for future study in the field of plant antioxidants, the applications and limitations of antioxidant activity measuring assays were also addressed.

3.4. Anti-Neurodegenerative Activity

Alkaloids, a structurally diverse category of secondary metabolites with high pharmacological effects, and nitrogen-containing secondary metabolites, which make up 60% of plant-derived medications, are among them. Alkaloids are a component of chemical defence in plants (Puri et al., 2018). Numerous botanical groups, including Papaveraceae, Asteraceae, Amaryllidaceae, Apocynaceae, Solanaceae, Fabaceae, Rubiaceae, and Rutaceae, are rich in alkaloids. Alkaloids exert a wide range of therapeutic effects due to their actions, including analgesic (morphine), antihypertensive (reserpine), antiasthmatic (ephedrine), anticancer (vincristine), antihyperglycemic (piperine), and antipyretic (quinine) effects. Thus, natural products have been used to treat a wide range of health issues since the dawn of time and have great therapeutic potential. For example, plant species with bioactive metabolites are effective treatments for neurological diseases, inflammation, and other associated health complications.

Modern medicine pays a lot of attention to plant-derived alkaloids in a regular supply of drugs to treat chronic illnesses like diabetes, neurological complications, and cancer. In order to support the ongoing quest for alkaloid-based therapies for both inflammatory and neurodegenerative illnesses, this chapter reviews and emphasises the particular role of alkaloids in modulating inflammatory and neurological pathologies (Liang et al., 2009). NDDs are typically seen in conditions like Alzheimer's disease, dementia, amyotrophic lateral sclerosis, and Parkinson's disease. Inhibitors of the enzymes monoamine oxidase (MAO), acetylcholinesterase, butyrylcholinesterase, and N-methyl-D-aspartate (NMDA), as well as agonists of the muscarinic and adenosine receptors, alkaloids have been proven to ameliorate the pathophysiology of NDDS. Numerous pieces of data suggest that traditional medicine formulations, which are primarily composed of plant-based ingredients, may be able to cure NDDs, allergic disorders, and inflammation with the least amount of systemic toxicity.

Alkaloids from plants exhibit anti-inflammatory properties by inhibiting a variety of pro-inflammatory protein complexes associated with inflammatory signalling pathways. This complex consists of inflammatory mediators such as cytokines, chemokines, prostaglandin E2 (PEG2), and nitric oxide (NO), as well as nuclear factor kappa -promote the activation of B cells (NF-kB), STAT1, extracellular signal-regulated protein kinase and Akt (Simmonds and Foxwell, 2008). An inflammatory state is identified by immune cell infiltration, activity, and the excessive synthesis of a number of cytokines and various inflammatory mediators (Guo et al., 2015). Inflammatory illnesses such as cardiovascular diseases, bowel disease, and rheumatoid arthritis are brought on by an excess production of inflammatory mediators, which also causes other diseases like diabetes, cancer, chronic kidney disease, neurodegenerative disorders (NDDs), and ageing.

Alkaloids may act as a neuroprotective agent by suppressing a variety of cellular processes, including the activity of the AChE enzyme, the production of GABA, an inhibitory neurotransmitter in the mammalian brain, the partial blockade of NMDA receptors, the stimulation of cellular autophagy, and many other processes. Neuron degeneration illness is characterized by ongoing, irreversible harm to the assembly or functioning of neurons as well as the accumulation of pathologically altered proteins in the human brain. Various neurodegenerative disorders (NDDs), including brain trauma, Alzheimer's disease, progressive supranuclear palsy, Parkinson's disease, amyotrophic lateral sclerosis, Huntington's disease, prion disorders, and spin cerebellar ataxias, may result from neuronal cell death (Aryal et al., 2022). Proteasomal dysfunction is a common mechanism underlying many NDDs, which causes the misfolded protein to be removed improperly and insufficiently, resulting in its aggregation in the brain (Trompetero et al., 2017). Additionally, the causes of NDD include oxidative stress, the generation of free radicals and reactive oxygen species (ROS), DNA repairs, mitochondrial dysfunction, and neuroinflammation.

In Alzheimer's disease, and fragmentation of neuronal cellular complexity, the most prevalent NDD, acetylcholinesterase (AChE) is associated with a plaque of extracellular amyloid protein (A) deposits and neurofibrillary tangles (Ng et al., 2015). The transmembrane-amyloid precursor protein (APP) is broken down by an enzyme called -site amyloid precursor protein cleaving the enzyme into a smaller polypeptide called A (BACE-1). Administered medications for the condition only work by suppressing AChE to lessen symptoms or stop disease development. Parkinson's disease, another prevalent NDD, has a similar environmental aetiology, genetic, and nongenetic. Misfolded protein aggregates, proteasomal dysfunction, neuroinflammation, mitochondrial DNA damage, oxidative stress, and genetic mutations are among the shared traits. Dopaminergic neuronal loss in the substantia nigra pars compacta and decreased dopamine levels are the distinguishing features of Parkinson's disease (Dauer and Przedborski, 2003). Multiple studies have discovered that post-mortem human brains of people with mixed dementia with Lewy bodies and Parkinson's illness with dementia commonly show the presence of numerous misfolded protein aggregates, including A, p-tau, and -synuclein. Monoamine oxidase-B (MAO-B), an enzyme that is elevated in the brain, catalyses the breakdown of dopamine, which results in lower levels of dopamine. This suggests that MAO-B is a good target to maintain dopamine levels in Parkinson's disorders (Raza et al., 2019). In order to treat neurodegenerative illnesses, plant-derived alkaloids have been used for centuries.

The body's innate system responds to both infectious (bacteria, viruses, fungi, and parasites) and noninfectious stimuli by inducing inflammation, which is a generalized and quick response mechanism. White blood cells such as dendritic cells, neutrophils, natural killer cells, monocytes/macrophages, eosinophils, and basophils are examples of innate immune cells. To maintain communication and coordinate immune responses, innate immune cells can create and release pro-inflammatory mediators like chemokines, hormones, NO, cytokines, adhesion molecules, and growth factors. Because they control genes involved in inflammatory and immune responses, nuclear factors, such as the NF-B family of inducible transcription factors, play a crucial role in the inflammatory process. A number of cytokines, cell adhesion molecules, chemokines, growth factors, and acute-phase proteins are all transcriptionally induced by NF-B (Aristizábal and González, 2013). The inhibitory protein of NF-B is phosphorylated and degraded by the proteasome as part of the activation process. As a result, the nucleus experiences a release of free NF-B, which then attaches to B binding sites in the promoters of target genes to trigger the transcription of pro-inflammatory cytokines. Tumour necrosis factor (TNF), interleukin (IL)-1, IL-12, IL-6, and IL-8 are some of the main cytokines it controls (Carroll et al., 2008). Additionally, it controls the expression of chemokines including CXCL-10 and C-X-C motif chemokine ligand (CXCL)-1, monocyte chemoattractant protein-1 (MCP-1), macrophage inflammatory protein-2 (MIP-2), and MIP-2. Inflammasomes, a multiprotein complex made up of pattern recognition receptors that serve as an innate immune system sensor to pathogenic microbes and host inflammatory proteins, are likewise modulated by NF B.

Overall, NF-B in innate immune cells is activated again by inflammatory cytokines and positive feedback, which leads to increased production of cytokines and chemokines and more infiltration of inflammatory cells to spread the inflammation (Liu et al., 2021). This response promotes the development and invasion of adaptive immune cells to get rid of pathogens and toxic antigens locally. However, deregulation of the inflammatory responses causes serious tissue damage and aids in the development of acute or chronic inflammatory disorders. Inflammation and the production of inflammatory mediators are often advantageous to the host as they aid in the resolution of diseases. In order to stop further tissue damage in the latter scenario, it is still advantageous to suppress inflammatory mediators or their receptors. Additionally crucial in triggering an inflammatory response are prostaglandins. They are the lipids that are biosynthesized in inflammatory and damaged tissue and are consequently linked to the emergence of acute inflammation symptoms including swelling and redness. Arachidonic acid, an important fatty acid, is converted into prostaglandins by cyclooxygenase (COX) isoenzymes. Nonsteroidal anti-inflammatory medications (NSAIDs) suppress COX activity and prevent prostaglandin formation (Gunaydin and Bilge, 2018). In addition to targeting prostaglandins, nitric oxide promotes and controls COX enzyme activity during the inflammatory condition. Since inflammation plays a role in the aetiology of many diseases, nitric oxide and/or COX are also frequently considered potential therapeutic targets.

3.5. Anti-Coronavirus Activity

Alkaloids are a group of naturally occurring nitrogen-containing chemicals with basic qualities and strong physiological effects. They typically have at least one heteroatom of nitrogen, typically in a heterocyclic ring. Alkaloids can be divided into three groups based on the biosynthetic pathway: (1) true alkaloids, which derive from amino acids and have a nitrogen-based heterocyclic ring; (2) proto-alkaloids, which also derive from amino acids but do not have a nitrogen moiety in a heterocyclic system; and (3) pseudo alkaloids, which have no relation to amino acids. Alkaloids are a class of natural chemicals that currently number over 8000. According to some estimates, 25 percent of Gymnosperms and Angiosperms generate alkaloids, which are extensively distributed across the plant kingdom (Dénès, 2018). Alkaloids are additionally found in a variety of plant groups, including Papaveraceae, Rutaceae, Apocynaceae, Asteraceae, Solanaceae, Fabaceae and Erythroxylaceae.

Alkaloids, flavonoids, terpenoids, lignins, and coumarins are only a few of the phytochemicals found in plants that have antioxidant properties and can block the viral genome. Alkaloids are a family of natural compounds with a wide range of pharmacological activity that holds considerable promise for the creation of novel therapeutics for a variety of diseases. Some alkaloids have antiviral properties and/or have served as prototypes for the creation of synthetic antiviral medications. Traditional uses of phytochemicals for disease treatment include reports that they can stop viral replication and transcription (Figure 5). The majority of them stop viruses either at the point of viral entry into the host cell or at the point of viral replication. The scientific literature was searched for eleven anti-coronavirus alkaloids in this study, and their potential therapeutic usefulness against the severe acute respiratory syndrome coronavirus-2 (SARS-CoV-2) was explored.

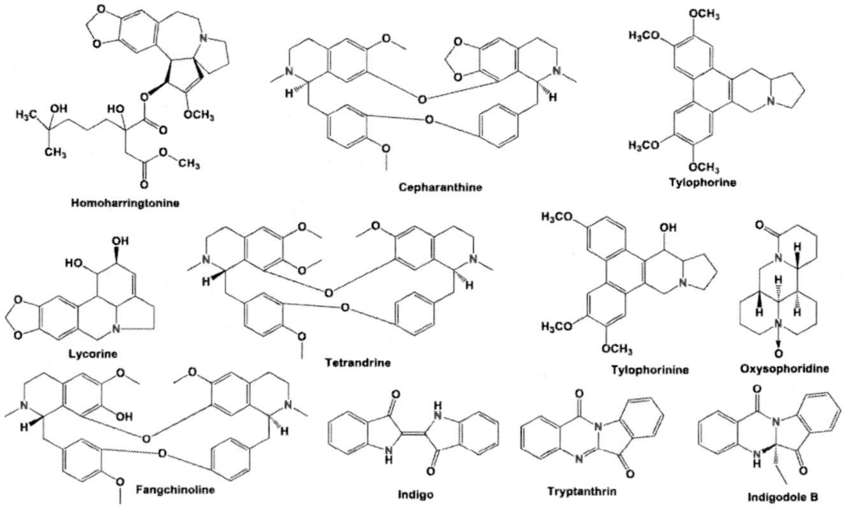

Figure 5. Chemical structures of different anti-coronavirus alkaloids (Fielding et al., 2020).

In silico experiments revealed an affinity of the alkaloids for binding to the receptor-binding region of the SARS-CoV-2 spike protein, ostensibly inhibiting its ability to interact with the host cell. Finally, many pathways for the alkaloids' known anti-coronavirus action were reviewed, demonstrating the alkaloids' potential as bioactive agents against SARS-CoV-2 (Majnooni et al., 2021). Numerous plant-derived products have been thoroughly investigated for their ability to combat viruses such hepatitis, herpes, and human immunodeficiency (HIV) viruses. More recently, a newly discovered coronavirus that causes Coronavirus disease (COVID-19) has spread globally and had a devastating impact on the population. However, phytochemicals for the suppression of viruses like dengue virus, chikungunya virus, and other alphaviruses are still understudied. The reported phytochemicals and their derivatives, which have antiviral characteristics and their mode of action to cure viral illnesses, will be the focus of this chapter.

On historical reports from several severe acute respiratory syndrome coronavirus (SARS-CoV) and Middle Eastern respiratory syndrome coronavirus (MERS-CoV) research, current pharmacological recommendations for the treatment of coronavirus disease-2019 (COVID-19) are based (Figure 6). Some data from research suggests that using an integrative strategy, such as combining western treatment with herbal remedies and/or natural substances derived from medicinal plants, is more successful in reducing the number of deaths and controlling coronavirus infection. Additionally, by lowering the effective concentration of substances below the therapeutic plasma level, successful combination therapy may improve clinical outcomes (Waterman, 1992). Alkaloids have been linked to a number of biological functions since they were first discovered, including analgesic, anti-inflammatory, anti-cancer, antibacterial, antifungal, and antiviral action. Berberine, tomatidine, michellamine B, oxymatrine, and palmatine are those alkaloids that showed higher antiviral activity. Berberine has demonstrated activity against the hepatitis C virus, chikungunya virus, and human cytomegalovirus (HCMV), while tomatidine has demonstrated activity against the dengue virus and HIV (ZV) (Panahi et al., 2013). The seeds of *Peganum harmala* L., which deactivate the impact of influenza A virus, and the root tubers of Stephania cepharantha Hayata, which increase the longevity of mice infected with herpes simplex virus type 1 (HSV1), are examples of plants having a higher concentration of alkaloids and exhibit antiviral activity.

The affinity of the eleven discovered alkaloids with anti-coronavirus action for binding to the receptor-binding domain of the SARS-CoV-2 spike protein was then investigated using in silico analysis. New antiviral treatments can be found in abundance in traditional herbal remedies and naturally occurring chemicals derived from plants. In actuality, around 25% of frequently prescribed medications contain isolated plant-based components. Numerous conventional herbal remedies include antiviral properties that are effective against a variety of viral strains. These properties affect the viral life cycle, including viral entrance, replication, assembly, and release, as well as the interactions between the virus and its unique hosts.

Animal and human coronaviruses have been reported to be broadly inhibited by alkaloids. Lycorine modifies host factors to prevent viral replication, in contrast to HHT, tylophorine, oxysophoridine, and tylophorine analogues, which inhibit CoV replication through unknown methods. Tetrandrine, Cepharanthine, and Fangchinoline exhibit virus translocation via the endolysosomal system or pointing viral RNA, hereafter preventing of TGEV replication happened. They also work in concert with a JAK-family inhibitor to provide comprehensive anti-CoV action. Conversely, Indigo prevents the 3CLpro from cleaving proteins in order to obstruct viral. Finally, tryptanthrin and indigodole B exhibit strong viricidal effects and reduce viral yield. In particular, it has been noted that tryptanthrin inhibits the activity of the papain-like protease 2 and viral RNA genome formation in HCoV-NL63-infected cells. As demonstrated by recent in vitro investigations and in silico simulations employing biological targets relevant to SARS-CoV-2, the alkaloids provide a promising potential as anti-coronavirus medicines and demand additional research.

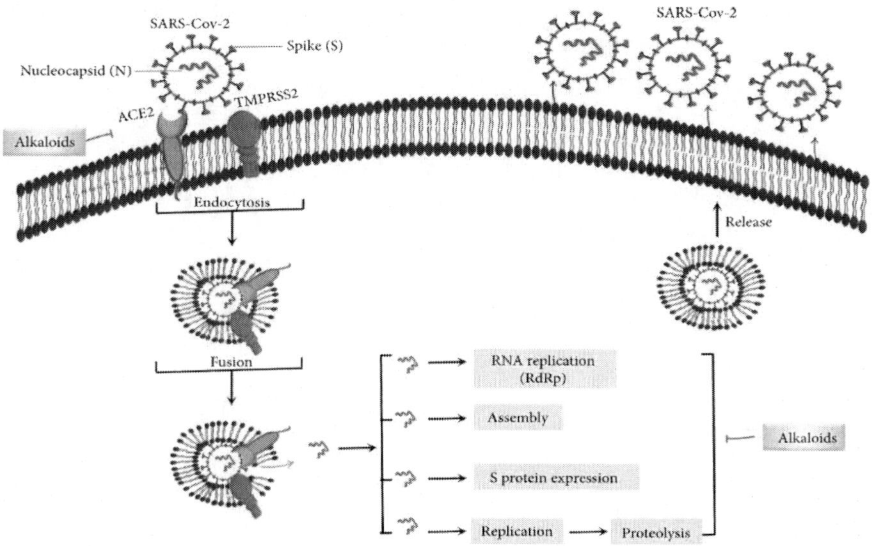

Figure 6. Alkaloids' primary targets in the fight against SARS-CoV-2 [79]

4. Utilization of Alkaloids in Current Medicine

Even while natural sources have been utilised as a source of medicine since at least 2600 BC and have significantly influenced the advancement of current medicine, many of the existing empirical rules do not take into account molecules with a variety of properties, notably in the case of natural products like alkaloids (Baker, 2021). Although many alkaloids are categorised according to their molecular structures, it is usually used to classify alkaloids according to the sources of their botanical origin for example lycodine-type alkaloids first obtained from the genus *Lycopodium* (Figure 7). Alkaloids provided distinctive lead compounds for therapeutic use. Alkaloids are lipid soluble in basic and neural settings and water soluble in acidic conditions as a result of their fundamental properties. This is essential for both the penetration of deprotonated membranes and the breakdown of protonated membranes. Alkaloids are essential for both human therapy and an organism's natural defence. Alkaloids form secondary metabolites found in plants. Plants have alkaloids that protect them from predators and regulate growth. The therapeutic uses of alkaloids as anaesthetics, anti-inflammatory medicines, and cardioprotectants are particularly well documented.

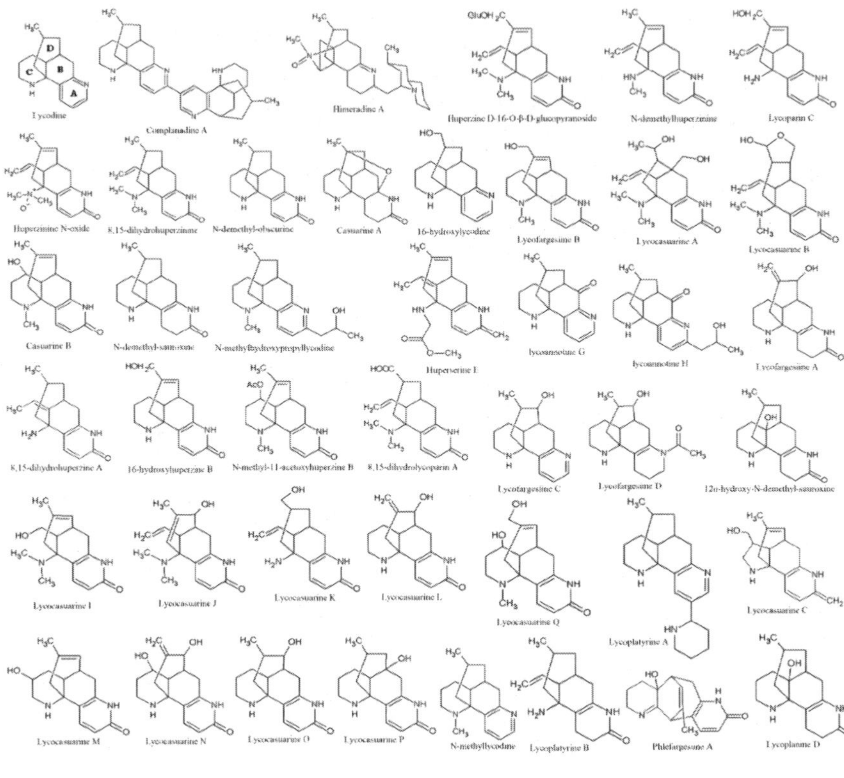

Figure 7. Chemical structures of selected lycodine type alkaloids (Patil, 2020).

Several well-known alkaloids that are used in clinical contexts include nicotine, quinine, strychnine, ephedrine, and quinine. Alkaloids have a wide range of applications. Including a variety of substances, such as poisons like caffeine, coniine, nicotine, and prescription drugs like quinine (Majnooni et al., 2021). It is important to keep in mind that many alkaloids utilised in medicine also have the potential to be toxic or addictive substances. Alkaloids are a paradox and a mystery. Strange, inexplicable, and yet essential to all existence. piperine, capsaicin, and the murraya alkaloids are a few examples of alkaloids we enjoy as seasonings; strychnine, aconitine, and batrachotoxin are a few alkaloids we despise as toxins; N,N-dimethyltryptamine, ibogaine, and psilocin are a few alkaloids that change perceptions in the brain; and caffeine, the stimulant found in coffee.

Alkaloids like morphine and cocaine, as well as others whose diverse effects can place a strain on healthcare systems, can both relieve pain and cause significant societal disturbance under certain circumstances (nicotine). Additionally, a number of alkaloids from other sources are employed as medicines (quinine, paclitaxel, atropine, pilocarpine, vincamine, vincristine, cephalosporins, and penicillin). Furthermore, natural compounds are more likely to imitate endogenous metabolites and metabolic intermediates than synthetic molecules that can be recognised as a substrate by active transporters.

Currently, very proactive developments in the field of traditional medicine (ethnopharmacology) have reignited interest in bioactive natural substances. The potential for these items to be utilised in the development of new drugs is another factor boosting interest in them. There were 27,683 alkaloids included in the Dictionary of Natural Products (DNP) as of October 25, 2020, and 990 newly discovered or previously studied alkaloids from nature were documented between 2014 and 2020 (Heinrich et al., 2021). The isolated drugs from the plants and other natural sources are needed to create a more specified medicine portfolio for biological applications and other uses. Despite improvements in discovery approaches, most notably the introduction of drugs produced from molecular biology, which has shown to be a highly successful tactic, the need for generating natural product-based treatments still continues.

A variety of chemical compounds, including alkaloids, which are frequently taken from plants, are produced through the biosynthesis of amino acids. A minor number of alkaloids are present in plant species, and research and development efforts are still heavily concentrated on increasing their production (including through biotechnology), extraction, and processing. For instance, altering the genetic code to boost the production of certain processes involved in the manufacturing of alkaloids (Changxing et al., 2020). Drug discovery approaches have changed significantly during the past few decades. The success of selecting drug development candidates using computational design and empirical approaches has been discussed extensively. Many studies are covering maximum chemical variety that is based on the concept of drug ability, which can be predicted as per pharmacokinetic parameters for leading compounds using hydrogen bond acceptor and donor theories (Madan and Dureja, 2012). The success of selecting drug development candidates using computational design and empirical approaches have been discussed extensively. Additional selection factors for leads include ligand efficiency, polar surface area, and rotatable bonds for absorption estimates. Since a wide range of potential lead molecules are excluded by the constraints of chemical synthesis, drug discovery screening libraries are anything but varied. A greater abundance of medicinal species with alkaloids than non-medicinal species is typically linked to cultivation. Simple alkaloids with high biodiversity are more likely to be used in the early stages of drug development than "rare" alkaloids since they are typically found in multiple species when "driving species" exist.

In collaboration with currently FDA-approved drugs, alkaloids may be employed to develop improved and long-lasting anti-inflammatory and anti-AChE formulations. The research of drug transport, medicinal chemistry, and ethnopharmacological uses of alkaloids in the treatment of inflammation and NDDs will all be significantly impacted by this chapter (Polis and Samson, 2018). Numerous excellent publications have been written about the challenges to be faced as well as the role of natural products in drug discovery. In particular, for overcoming multiple drug resistance and for treating uncommon and tropical diseases, such as neglected tropical diseases, the focus will be on how alkaloids act in the complex environment of the continuing discovery process. It will require a creative approach to the question of how the current discovery as well as existing and expected levels of technical approaches may enhance the use of alkaloids as pharmaceuticals in novel ways in order to meet these needs for the welfare of the patient and the world's health. What paradigms in the methods used in discovery programmes need to change if alkaloids are to continue to be a useful class of medicines?

Alkaloids are among the most intriguing categories of natural chemicals. For unique potential preventative and/or therapeutic uses in AChE inhibition, NDDs, and anti-inflammatory, they provide a wide spectrum of structurally and/or functionally varied compounds. A literature study and in silico analysis revealed that a number of alkaloids, including oxymatrine, tetrandrine, tetrahydropalmatine, harmine, berberine, aloperine, and sinomenine, have the ability to act as lead compounds against various anti-inflammation and NDDs (Batool et al., 2021). It must be carefully examined for clinical trials, pharmacokinetic qualities, health issues, and other critical factors before being used as a treatment. It is crucial to carefully assess the safety profiles of alkaloids because many of them are harmful as well (Monte et al., 2014). *Aegle marmelos* showed antioxidant qualities that reportedly guard against a variety of free radicals, according to sources. It was reported here that unripe fruit exhibited a larger percentage of free radical inhibition than ripe fruit. Bacillus subtilis had the highest level of antibacterial activity, followed by *E. coli, Pseudomonas aeruginos* and *Staphylococcus aureus* (West et al., 2014). Numerous animal and human fungi, including *Trichophyton mentagrophytes,*

Epidermophyton floccosum, Trichophyton rubrum, Aspergillus niger, Microsporum gypseum, Histoplasma capsulatum, Aspergillus flavus, Microsporum audounii and *Microsporum cookie* have been shown to be resistant to the antifungal effects of the essential oil that was taken out from the leaves.

According to a different study, the bilwa methanol extract demonstrated strong antibacterial effects against *Salmonella paratyphi* A and B, *Klebsiella pneumonia, Escherichia coli, Staphylococcus aureus,* and *Bacillus subtilis.* The agar well diffusion method was also used to test the antibacterial activity of different extracts. Hexane, ciprofloxacin extracts, cold and hot methanol were used to demonstrate high antibacterial activity against *Streptococcus faecalis, Proteus vulgaris, Micrococcus luteus, Escherichia coli, Klebsiella pneumonia*, and *Enterococcus faecalis*. In *Aegle marmelos*, antidiarrheal qualities have been reported in numerous investigations. *Aegle marmelos* has been found to have antidiabetic qualities in various studies, hence it has this property. *A. marmelos* leaves were discovered to possess anti-diabetic effects in rats with alloxan-induced diabetes (Aryal et al., 2021a). The methanolic extract of *A. marmelos* leaf lowers blood sugar levels. According to this, it was found that continued administration of the extract caused a drop in blood sugar levels, and after 12 days, it was found that the sugar level had fallen by 54% Leaf extract has been used to treat diabetes in ayurvedic medicine. It increases the body's ability to utilise the external glucose load by stimulating glucose absorption through the utilisation of insulin.

Cancer is the second-leading cause of death for both men and women worldwide, both in industrialised and developing countries. A body's anticancer activity is increased by using the *Aegle marmelos* fruit extract to boost the immune system (Iqbal et al., 2017). According to a study, the *A. marmelos* has an anticancer effect in an animal model with carcinoma (Aung et al., 2017). Numerous studies have demonstrated the anti-cancer benefits of the phytochemicals present in *A. marmelos*, such as lupeol, d-limonene, eugenol, and citral, cineole. Modern medicine is not as safe to use as natural medicine when it comes to antipyretic activity. *A. marmelos* is also used to treat pain and fever since it contains antipyretic qualities.

5. Promising Plant-Derived Alkaloids

Traditional medicines and natural goods are very important. In various regions of the world, traditional medical practises including traditional Chinese medicine, Ayurveda, Kampo, Unani, and traditional Korean medicine have developed into systematic, well-organized medical systems. When used to create novel pharmaceuticals, natural products and conventional medicines have a number of advantages, including a wealth of clinical experience, a wide variety of chemical structures, and biological activity (Waterman, 1992). According to research, various pyrrolidine alkaloids have been demonstrated to possess a range of significant biological properties, including antibacterial, anticancer, anti-hyperglycemic, antifungal, antioxidant, anti-inflammatory, antiparasitic, anthelmintic, organ protecting, and neuropharmacological properties. Additionally, it has been discovered that several alkaloids have harmful effects on the organs of animals. While nicotine and cocaine have been proven to produce neurotoxicity in test animals, bgugaine, irniine, and alkaloids are known to cause kidney damage. Additionally, some of the finest sources of lead compounds with pharmacological activity might come from pyrrolidine alkaloids.

5.1. Tetrahydropalmatine

The isoquinoline alkaloid tetrahydropalmatine (THP), which is mostly found in the Stephania and Corydalis genera, exhibits anxiolytic, anti-inflammatory, analgesic, and cardioprotective properties. Tetrahydropalmatine's impact on the human monocytic cell line THP-1's response to lipopolysaccharide (LPS)-induced interleukin (IL)-8 production was investigated to shed light on the substance's biological effects. Reverse transcription-polymerase chain reaction, enzyme-linked immunosorbent assay, mitogen-activated protein kinase (MAPK) activation, and western blot analysis were explored to evaluate IL-8 synthesis in the current investigation (Xue et al., 2008). Tetrahydropalmatine reduced the amount of IL-8 produced and also prevented the phosphorylation of p38 mitogen-activated protein kinase and extracellular signal-regulated kinase, suggesting that it prevents the release of IL-8 via preventing MAPK phosphorylation. Together, these results may shed light on how tetrahydropalmatine influences THP-1 cell activation in an inflammatory environment (Gao et al., 2016). The protective effect of THP against nerve cell apoptosis has been reported by attenuating the ketamine-induced surge in AChE activity and overriding the ketamine-induced reduction in ACh levels.

According to studies, THP therapy improved memory deficits brought on by D-galactose by increasing glutathione levels, superoxide dismutase (SOD), catalase, and glutathione peroxidase activities while decreasing levels of malondialdehyde (MDA) and NO. Similar to this, THP administration had a protective effect on animals against ketamine-induced oxidative stress by raising glutathione peroxidase and SOD activity while lowering MDA activity. Additionally, THP reduced the expression of TNF-, IL-1, and IL-6, inhibited the activity of the proteins iNOS and NF-kB, and increased the production of the protein glial cell-derived neurotrophic factor in animals given ketamine to promote inflammation (He et al., 2011). Inhibiting both the mRNA and protein levels of vascular cell adhesion molecule-1 (VCAM-1) as well as the attenuation of TNF-stimulated NF-B translocation in monocytes, l-THP inhibited TNF-induced adhesion of monocytes to human umbilical vein endothelial cells, highlighting its potential pharmacological action to prevent atherosclerosis (Poddar et al., 2021). Therefore, in-depth pharmacological research should be done in the future to explain its anti-inflammatory and neuroprotective effects.

5.2. Aloperine

Aloperine, a piperidine alkaloid, has anti-inflammatory and anti-neuropathic effects. It is obtained from the plant *Sophora alopecuroides*, which is widely used as medicine throughout Central and Western Asia. According to the reductionist philosophy, later biochemical tests link this anti-dysentery action to aloperine's bactericidal properties. After that, aloperine's various functions start to become more apparent. However, the existing knowledge on aloperine is fragmented and needs to be summarised. Growing evidence reveals that aloperine has diverse pharmacological properties and bears great potential in clinical situations like skin hypersensitivity, tumour and inflammatory disorders, etc. (D'Orazio et al., 2012; Zhou et al., 2020).

Aloperine's anti-inflammatory properties were established by its capacity to inhibit the macrophage Toll-like receptor 4 (TLR4)-dependent inflammatory pathway. Thus, it was demonstrated to inhibit COX-2 and iNOS, thereby reducing PEG2 secretion and blocking the expression of IL-17A, TNF-, and IL-6 (Xu et al., 2022). This in turn decreased NO generation. Similar to this, treatment with aloperine decreased oxidised low-density lipoprotein, a sign of endothelial inflammation, as well as IL-6, and E-selection by lowering the expression of Kruppel-like factor 2 (KLF2), indicating possible anti-atherosclerosis properties. More proof demonstrates that aloperine reduced neuropathic pain brought on by chronic constriction injury in the dorsal spinal cord. This was accomplished by preventing the upregulation of IL-1, NF-B, and IL-6, which is connected to the lowering of ROS through the suppression of NF-B pathways (Skundric et al., 1997).

5.3. Tetrandrine

Tetrandrine, a bisbenzylisoquinoline alkaloid widely used to treat inflammation, was discovered in the roots of *Stephania japonica*, S. Moore, and *S. tetrandra*. One sort of natural substance, tetrandrine, was first made from Chinese botanicals. The bisbenzylisoquinoline alkaloid tetrandrine [(1b)-6,6′,7,12-tetramethoxy-2,2′-dimethyl-berbaman] has a number of pharmacologically significant classes. Tetrandrine had been used as a therapeutic agent in China for treating inflammatory lung diseases, silicosis, autoimmune conditions, cardiovascular ailments, and hypertension. Tetrandrine offers pharmacological potential for cancer therapy, according to several early investigations. The development of aloperine as a medication to treat many illnesses, including inflammation and neurological diseases, requires more investigation despite the fact that it has a wide variety of medical applications. By inhibiting NF-kB signalling in LPS-induced macrophages, tetrandrine has been found to reduce the release of pro-inflammatory mediators, IL-6, TNF-, and IL-1 expression (Xu et al., 2014). Tetrandrine also prevents IkB phosphorylation in ATDC5 cells and LPS-induced cells, which in turn prevents the expression of matrix metalloproteinase-3, tissue metalloproteinase inhibitor-1, the generation of PEG2 (prostaglandin E2) and NO (nitrite oxide). Tetrandrine also decreased the levels of NO in the serum and pancreatic tissue of acute hemorrhagic necrotizing pancreatitis-prone rats, and it blocked NF-B activation by focusing on the production of IL-8, TNF-, and IL-6 (Borthakur et al., 2010). Tetrandrine was also administered intravenously to rats with an Alzheimer's disease model, and these results included improvements in memory and learning impairment as well as a reduction in TNF- and IL-1 levels thanks to the inhibition of the NF-B pathway. It's interesting to note that tetrandrine pills, or tetrandrine, have recently been utilised in a phase 4 clinical trial to treat COVID-19 patients with anti-inflammatory medication. Overall, the evidence shows that tetrandrine is an effective alkaloid for treating neurological illness and inflammation.

Tetrandrine has the most advantageous effects on tumour cells through the reduction of proliferation and activation of apoptosis, not only on cancer cell lines but also on primary cancer cells separated from ascites and pleural fluids, which were obtained from individuals with stomach, colon, liver, and lung tumours. Tetrandrine also reverses multidrug resistance (MDR), inhibits angiogenesis, radiosensitizes tumour cells to radiation, and metastasis, among other biological effects. Publications from that era did not, however, explicitly explore the mechanisms behind these impacts. Tetrandrine's anti-tumour bioactivity has recently been successfully demonstrated in a number of investigations (Dholwani et al., 2008). Numerous research especially looked into the ways in which tetrandrine treats different cancer cells. Tetrandrine's ability to induce autophagy and a strategy for effectively delivering it to cancer cells have just been discovered (Yuan et al., 2016). Additionally, mounting data suggests creating nanoscale delivery systems to carry tetrandrine into cancer cells in order to increase the anti-tumour effects of the drug. Notably, other derivatives of the bisbenzylisoquinoline alkaloid, like fangchinoline, also have an impact on cancer cells.

5.4. Sinomenine

Sinomenine, a type of benzyl alkaloid frequently used in Chinese herbal medicine, is mostly extracted from *Sinomenium acutum* root and stem. As a powerful histamine releaser in conjunction with tissue mast cell degranulation in mammalian tissues, Sinomenine is a special plant alkaloid. The skin and joint capsules are where this action occurs most frequently. The primary pharmacological effects of sinomenine, including stimulation of gastric acid secretion, acceleration of thoracic, vasodilatation, increased vascular permeability, contraction of smooth muscles, peripheral lymph flow, and increased peristalsis of the intestines are caused by histamine that has been released (Yamasaki, 1976). Convulsive central excitation was seen in the majority of laboratory animals at hazardous levels of sinomenine. Clinical adverse effects included injection site inflammation, itching in the head and upper body, oedema around the lips and eyelids, and transient cephalalgia when sinomenine or *Sinomenium acutum* were administered in high dosages. Classical antihistamines significantly minimised the majority of these negative effects (H1-receptor antagonists). Simomenine was injected under the skin every day for over a week, and this had an analgesic impact on mice. Sinomenine hydrochloride or histamine dihydrochloride prevents the formation of granulation tissue and adjuvant arthritis. Since both effects were obviously suppressed by the H2-antagonist burimamide but not by the H1-antagonist mepyramine, these inhibitory effects were most likely mediated through histamine H2-receptors on T-cells (adjuvant arthritis) as well as fibroblasts (granulation tissue growth) (Levi et al., 1975). Anti-rheumatic effect of *Sinomenium acutum* is undoubtedly real, and it can be explained by the histamine-releasing abilities of sinomenine.

Chronic nephritis, rheumatoid arthritis, myocardial ischemia, ankylosing spondylitis, and other fast arrhythmias can all be treated with it therapeutically. It also lessens any accompanying foot swelling brought on by formaldehyde, egg whites, or carrageenan. Inhibiting c-Jun N-terminal kinases (JNKs) and NF-B signalling pathways prevents cytokines like IL-1 and TNF- from expressing their mRNA, which is how sinomenine exerts its anti-inflammatory effects (Koch et al., 2015). Simomenine also has the potential to be an effective treatment for NDDs. It has been demonstrated to suppress ROS and NO production in human astrocytes treated with A, suggesting a beneficial impact on Alzheimer's disease. In the rat model of temporal lobe epilepsy in intrahippocampal kainate, it also accounted for considerable neuroprotective potential by reducing seizure frequency, severity, status epilepticus occurrence, aberrant mossy fibre sprouting (MFS), and DNA fragmentation. Sinomenine and methotrexate were used as drugs to treat rheumatoid arthritis, finally, they improved it by lowering the expression of pro-inflammatory cytokines. Thus, sinomenine is a useful alkaloid for treating both inflammation and neurodegenerative illness.

5.5. Oxymatrine

Since many years ago, *Sophora flavescens* alkaloid (SFA) gels, a substance used in Traditional Chinese Medicine and also known as Kushen, have been used in clinical settings in China. Numerous clinical studies have shown that SFA gels have an anti-cancer effect. According to this research, SFA gels can be used to treat a variety of cancers, including those of the lung, breast, stomach, liver, ovary, and colorectal kinds. According to Chinese researchers, SFA gels can be used to treat cervical erosion, vaginal fungal infection, and aerobic vaginitis. Its principal active constituents are matrine and oxymatrine, and mounting research has shown that these substances have anticancer effects on a variety of cancer types, including acute myeloid leukaemia, prostate cancer, lung cancer, and others (Liu et al., 2014). For matrine, it can prevent the growth of cancer cells by preventing cell division, preventing cell death and cell cycle arrest, preventing invasion, and controlling specific signalling pathways. Additionally, several research studies suggested that the antiproliferative, antioxidant, and anti-inflammatory properties of oxymatrine may help to avoid pulmonary hypertension.

Human prostate and pancreatic cancer cells can undergo apoptotic cell death when exposed to oxymatrine. According to certain studies, the combination of oxymatrine and HMGB1 may be more successful in treating human synovial sarcoma, and oxymatrine may also induce autophagy. These investigations suggested that the anti-tumour properties of matrine and oxymatrine may be useful. It has not yet been determined how SFA gels affect cervical cancer cells or how they work (Figure 8). Oxymatrine, a quinolizidine alkaloid that may be obtained from the roots of *Sophora flavescentis*, has anti-inflammatory, anti-cancer, cardiovascular protective, anti-allergic, antiviral, and anti-fibrotic properties (Runtao et al., 2011).

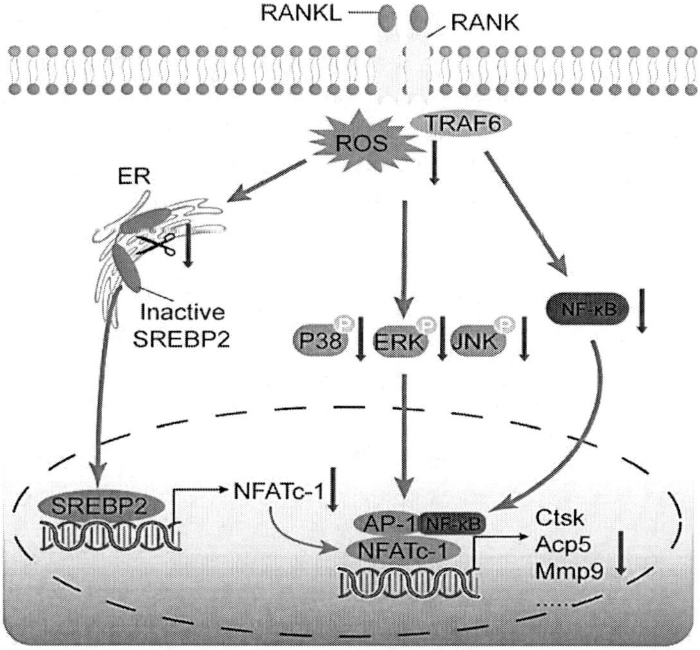

Figure 8. A proposed scheme for the mechanism by which OMT attenuates osteoclastogenesis via modulation of ROS-mediated SREBP2 Signaling. RANKL stimulation could induce ROS generation and activate downstream MAPK and NF-κB pathways (Jiang et al., 2021).

Oxymatrine has been demonstrated to have a strong anti-inflammatory effect on collagen-induced arthritis (CIA) rat models of inflammation because it decreased the production of cytokines, TNF- and IL-17A, as well as the arthritic score and synovial inflammation. Through the control of inflammatory responses in the stomach, pro-apoptotic effects, reduction of oxidative stress, and oxymatrine have extraordinary preventive benefits on gastric ulcers. By preventing NF-B from moving from the cytoplasm to the nucleus, it was discovered to stop a number of inflammatory mediators from entering ulcerated tissue (Trask, 2004). Additionally, oxymatrine has been shown to have anti-disease Alzheimer's benefits by reducing the density of A plaques and astrocyte clusters and enhancing learning and cognitive function in a mouse model of the condition. Oxymatrine dramatically lowers 1-methyl-4-phenyl-1,2,3,6-tetrahydropyridine-induced Parkinson's disease, resulting in dopamine-based neuroprotection and a decrease in microglia-mediated neuroinflammation through cathepsin-D-dependent modulation of the Toll-like receptor 4 (TLR4) signalling pathway (Bhardwaj and Deshmukh, 2018). The clinical trial proved that oxymatrine treats psoriasis and performed well as an effective anti-inflammatory and neurodegenerative treatment. It also had effective neuroprotection against cerebral hypoxic-ischemic injury by reducing apoptosis and oxidative stress, which may be related to activating protein kinase (Akt) and glycogen synthase kinase-3 (GSK3) and changing factor 2/heme oxygenase 1 (Nrf-2/HO-1) signalling pathway.

5.6. Galantamine

A common source of the plant alkaloid galantamine is the Amaryllidaceae family, which includes the species *Galanthus woronowii* and *Leucojum aestivum*, used for treating Alzheimer's disease. It is now a significant therapeutic alternative utilised to delay the neurological deterioration process in AD. Galantamine was widely used as a recognised treatment in many Eastern European nations for Myasthenia gravis and muscular dystrophy, poliomyelitis residual symptoms, trigeminal neurologica, and other types of neuritis based on knowledge of galantamine's effects on the peripheral and central nervous systems (Wilkinson, 2001). Acetylcholine esterase showed inhibitory effects of galantamine isolated from *Galanthus woronowii*. Studies showing that the alkaloid crosses the blood-brain barrier are what first led to the development of CNS-related symptoms. Galantamine's ability to treat AD was greatly aided by a synthesis created in the middle of the 1990s. The cholinergic hypothesis serves as the basis for the scientific case for utilising cholinesterase inhibitors in the treatment of AD. One defining feature of AD is the impairment of the central cholinergic system, which is indicated by the death of cholinergic neurons in the forebrain and a significant decline in the activity of choline acetyltransferase. Galantamine serves as an illustration of how a natural product can be successfully transformed into a clinically significant drug through the use of ethnobotany.

Galantamine has been identified as a promising treatment for neuroinflammation and cognitive impairment in neurodegenerative illnesses due to its ability to inhibit cytokines (IL6, IL-1, and TNF-), gliosis, and pro-inflammatory signalling molecules (NF-B p65). These effects were observed in mice exposed to lipopolysaccharide (Liu et al., 2018). Additionally, it demonstrated neuroprotective activity via the PI3K-Akt and Bcl signal transduction cascade at nicotinic receptors. Galantamine increased the rat cortical neurons' NMDA responses, indicating its significance in improving learning, memory, and cognition in people with Alzheimer's disease. Acute treatment has been demonstrated to increase the levels of hippocampus insulin-like growth factor 2 mRNA in mice, indicating a neurogenetic effect. Clinical trials for galantamine hydrobromide have been done for treating dementia, mental illnesses, Alzheimer's, and brain ailments and showed remarkable results (Wilkinson, 2001). The incidence of adverse effects, especially those involving the gastrointestinal system and mediated cholinergically, is typically minimal and can be reduced by following the suggested cautious dose-escalation method. Therefore, galantamine might contribute to lessening the overall difficulty and expense of caring for dementia sufferers. Galantamine could eventually replace other therapies for dementia, according to the available data.

5.7. Berberine

In Indian and Chinese medicine, berberine an isoquinoline alkaloid of the protoberberine type found in a variety of plants has been employed as an antibiotic, stomachic, bitter tonic, and in the management of oriental sores (Casciaro et al., 2020). These herbs have been used for generations in traditional Chinese medicine to cure diabetes without medication. Additionally, there is a long history of using this substance to treat diarrhoea, gastroenteritis caused by bacteria, and other digestive illnesses. Alkaloids are a type of organic molecule that is mostly composed of basic nitrogen atoms and are found in plants. They can have significant physiological effects on people, particularly with regard to cardiovascular and metabolic health (Roy et al., 2021). Although berberine has been the subject of numerous pharmacological studies in the past, there is currently increased interest in the drug due to its alleged positive effects on a number of neurodegenerative and neuropsychiatric illnesses. According to reports, the alkaloid alters the brain's neurotransmitter and receptor systems.

Berberine chloride reduced hippocampal neurodegeneration and lowered BACE-1 activity in a rat model of Alzheimer's disease. Berberine dramatically decreased plaque aggregation in the transgenic mice model of Alzheimer's disease, which improved neural and mental dysfunction via inhibiting APP phosphorylation. The goal of this chapter is also to explore the pharmacological basis for berberine's usage in treating different central nervous system illnesses and associated conditions. In mental depression, Alzheimer's, schizophrenia, anxiety, and cerebral ischemia, its preventive effects are underlined.

For berberine to be used as a treatment for neurological illnesses, more thorough clinical studies and a safety evaluation are necessary. Benzylisoquinoline alkaloid (BIA) class bioactive chemical berberine is present in the medicinal plant *Berberis koreana* (Rubio-Pina and Vazquez-Flota, 2013). For its advantages to health, BIA is frequently employed in the food and pharmaceutical industries. At various growth phases, investigation of the berberine production pathway and gene expression has been carried out on leaves, flowers, and fruits. The absence of genes for the enzymes of other BIAs but the inclusion of all the genes for berberine biosynthesis in *B. koreana* explain the simplification of berberine biosynthesis. The genes encoding enzymes for other BIAs, other than those encoding corytuberine synthase (CTS) and berbamunine synthase, were absent from our dataset in addition to those for the berberine biosynthetic pathway (Durairajan et al., 2012). This explains how *B. koreana* synthesises berberine by effectively inhibiting the pathways leading to other BIAs and allowing the pathway to just lead to berberine synthesis. According to a growing body of research, berberine benefits may include defence against conditions such as metabolic syndrome, heart disease, high cholesterol, diabetes, gastroenteritis, hypertension (high blood pressure), joint issues, low bone density, immune challenges, weight management, cognitive decline, and possibly depression (Kim et al., 2014). By activating AMP-activated protein kinase (AMPK) in macrophages and inhibiting the production of pro-inflammatory genes such as TNF-, IL-I, IL-6, monocyte chemoattractant protein-1 (MCP-1), COX-2, and iNOS, berberine has an anti-inflammatory effect.

Most importantly, berberine also shows an anti-inflammatory effect in hepatocytes by preventing TNF- and IL-6 production in HepG2 cells. The usage of berberine in the NDDs model has received extensive research. Currently, berberine is undergoing various stages of clinical trials to treat a number of illnesses, such as hypercholesterolemia, atherosclerosis, schizophrenia, Alzheimer's disease, and coronary artery disease. In order to effectively treat inflammatory and neurodegenerative illnesses, berberine must be a significant medicine.

5.8. Harmine

Beta-carboline alkaloid harmine is found in large quantities in plants, animals, human tissues, insects, mammals, and bodily fluids. Harmine has several different kinds of pharmacological effects, including anticancer, cytotoxic, antiplasmodial, antibacterial, antifungal, antioxidant, antigenotoxic, antimutagenic, and hallucinogenic ones (Patel et al., 2012). It improves insulin sensitivity and has vasorelaxant effects via acting on the monoamine oxidase A or B receptor and gamma-aminobutyric acid type A receptor. Harmine stops osteoclastogenesis, which stops bone loss. The current section provides an overview of the pharmacological activity and analytical methods of harmine, which may be helpful for researchers to investigate the substance's untapped potential and create novel medications for treating a variety of illnesses.

The -carboline alkaloid harmine is highly significant in terms of pharmacology due to its anti-inflammatory, psychedelic, antifungal, antibacterial, antioxidant, and anticancer properties. *Peganum harmala* seeds were initially employed for their isolation (Huang et al., 2022). Harmine's anti-inflammatory properties were proven in mice that had received LPS injections. It was discovered to block NF-B activation, which decreased the levels of IL-1, TNF-, and IL-6 in the serum. Inflammatory factors like IL-1, NO, and TNF may be inhibited by harmine-loaded ethosomes, which are useful for reducing inflammation in a rat paw oedema brought on by carrageenan. In addition, it increased the activities of SOD while down-regulating the expression of TLR4, the nucleotide-binding oligomerization domain (NOD), pyrin domain-containing protein 3 (NLRP3), and leucine-rich repeat (LRR), as well as myeloperoxidase activity and MDA generation. Scopolamine-induced inflammation was decreased as a result of decreased TNF-, myeloperoxidase, and NO production activity (Richardson et al., 2010). It suppressed NLRP3 inflammasome activation by lowering NLRP3, apoptosis-associated speck-like protein containing a caspase-recruitment domain (ASC), cleaved In scopolamine-induced mice, harmine improves memory impairment by boosting cholinergic function through the inhibition of AChE, indicating that harmine may be an effective treatment for neurodegenerative illness. As a result, harmine may be considered a powerful NDD and anti-inflammatory medication.

6. Biological Roles and Bioanalytical Aspects

Many plants having alkaloids as key components have been used for medicinal purposes in traditional medicine. Selected examples of alkaloid-rich plants and their biological activities are shown in Table 1.

Table 1. Various alkaloids from different plants and their applications in traditional medicine

S.No.	Plant/Family	Alkaloids present	Biological activity	Ref.
1	*Adhatoda vasica* Acanthaceae	vasicine; vasicinol; vasicinone; adhatodine; vasicinolone; anisotine; peganine	antibacterial, antiulcer, antitubercular, antiallergic, respiratory conditions, rheumatic pain	Gangwar and Ghosh, 2014
2	*Arisarum vulgarae* Aracaceae	bgugaine; irniine	antibacterial, antifungal	Melhaoui et al., 1993
3	*Bleekeria vitensis* Apocynaceae	ellipticine; 9-methoxyellipticine; isoreserpiline; carapanaubine	anticancer	Sainsbury and Webb, 1972
4	*Casimiroa edulis* Rutaceae	edulein; scopoletin methyl ether; zapoterin; casimiroedine; isoimpinellin	anticancer, antiepileptic, anticonvulsant, antihypertensive	Awaad et al., 2007
5	*Cocculus hirsutus* Menispermaceae	shaheenine; cohirsinine; hirsutine; jamtinine; cohirsine; cohirsitine; haiderine; isotriboline; d-triboline; dl-coclaurine	diuretic, analgesic, antidiabetic, antiinflammatory, antimalarial, cardiotonic, antimicrobial	Bothara et al., 2011
6	*Cynanchum paniculatum* Apocynaceae	antofine	antitumor, antiinflammatory, neuroprotective	Gao et al., 2017

S.No.	Plant/Family	Alkaloids present	Biological activity	Ref.
7	*Evodia rutaecarpa* Rutaceae	evodiamine; rutaecarpine; dehydroevodiamine; evodiakine; 2-undecyl-4(1h)-quinolone	antitumor, anticancer, uterotonic, antiinflammatory, antinociceptive	Lu et al., 2020
8	*Hernandia nymphaeifolia* Hernandiaceae	S-ovigerine; S-N-methyl ovigerine; S-magnoflorine; shernovine; S-N-hydroxyovigerine	antiplasmodial	Roy and Chowdhury, 2015
9	*Lindera megaphylla* Lauraceae	dicentrine; dicentrinone; reticuline; northalifoline; nmethyl nandigerine	antiplatelet activity	Chen et al., 2016
10	*Nauclea orientalis* Rubiaceae	naucleficine; naucleactonine; naucleaorals A & B	antimalarial	Sichaem et al., 2018
11	*Peganum nigellastrum* Nitrariceae	luotonins C & D; harmine; 3-phenylquinoline; pegamine β-dglucopyranoside; 2-deoxypeganylacetic acid	antileishmanial	Ma et al., 2000
12	*Rauwolfia vomitoria* Apocynaceae	sarpagan; picrinine; akuammiline; heteroyohimbine; yohimbine; aricine; isoreserpiline; rauvoxine; rauvoxinine	antimalarial, anti-psychotropic, coughs, skin infections	Erasto et al., 2011
13	*Solanum khasianum* Solanaceae	solasonine; solamargine; khasianine	anti-inflammatory, anthelmintic	Sahu and Mahato, 1994
14	*Uncaria tomentosa* Rubiaceae	pterodine; isomitraphylline; uncarine F; mitraphylline; isopteropodine	antiinflamamtory, antineoplastic, antimicrobial, antioxidant	Bacher et al., 2006
15	*Aegle marmelos* Rutaceae	aegeline; marmeline; shahidine; skimmianine; ethylcinnamamide	antiinflammatory, antimicro, anticancer, antioxidant	Lambole et al., 2010
16	*Aristolachia manshuriensis* Aristolochiaceae	manshurienine A & B	antiinflammatory, diuretic, antibacterial, antineoplastic	Zhang and Jiang, 2006
17	*Cassytha filiformis* Lauraceae	actinodaphnine; cassythine; dicentrine	anticancer, antioxidant, diuretic	Mythili et al., 2011
18	*Cocculus laurifolius* Menispermaceae	isococculidine; isococculine; coccuvine; stepharine	diuretic, vermifuge, muscle relaxant	Marya and Bothara, 2011

Table 1. (Continued)

S.No.	Plant/Family	Alkaloids present	Biological activity	Ref.
19	*Cynanchum vincetoxicum* Apocynaceae	13-alpha antofine; 13-alpha-6-O-desmethyl antofine	anticancer	Staerk et al., 2002
20	*Ficus benjamina* Moraceae	salsoline; reticuline; anabasine; matridine; ajamalicine; columbamine; laudanosoline; isoclaurine	anti-inflammatory, antimalarial, antimicrobial, antipyretic, antinociceptive	Novelli et al., 2014
21	*Hippeastrum puniceum* Amaryllidaceae	3-O-acetyl-narcissidine, lycorine, pancratistatin, nasciclasine	antitumor, antiviral, antimalarial	Bribi et al., 2015
22	*Litsea glutinosa* Lauraceae	boldine; litseglutine A & B; laurolitsine	antidiarrhoeal, antidysentery,	Yang et al., 2005
23	*Nicotiana glauca* Solanaceae	anabasine; nornicotine	antibacterial	Abdel Rahman et al., 2011
24	*Physalis minima* Solanaceae	withaminimim; phygrine	diuretic, analgesic, antipyretic, antimalarial, antiulcer	Chothani and Vaghasiya, 2012
25	*Rhizoma coptidis* Ranunculaceae	berberine, palmatine, epiberberine, coptisine, jatrorrhizine, magnoflorine	antihypertension, antihyperglycemia, anti-inflammatory	Tan et al., 2016
26	*Solanum nigrum* Solanaceae	solasodine	anti-inflammatory, diuretic, antipyretic, antibacterial, antitumor	Atanu et al., 2011
27	*Tabernaem divertica* Apocynaceae	cononitarine B; conophylline	anti-inflammatory, antimalarial, antidiarrhoeal, antidysentery, analgesic, antipyretic	Ikewuchi et al., 2015
28	*Voacanga africana* Apocynaceae	voacafrine; voacafricine; voacamine; vobtusine; vobasine; voacangine; epiibogaine; voacangine hydroxyindolenine	anti-inflammatory, anti-gonorrhoel, antifungal	Mei et al., 2012
29	*Aframomun meleguata* Zingiberaceae	10,12-dihydroxy-18-ethenyl-4-pyrido-β-carboline	anti-inflammatory, antioxidant, antidiabetic, antimicrobial, neuroprotective	Gangwar and Ghosh, 2014

S.No.	Plant/Family	Alkaloids present	Biological activity	Ref.
30	*Altonia angustiloba* Apocynaceae	yohimbine; cathafoline; cabucraline; vincamajine; normacusine B; lochnerine; alstophylline; acralstonine; villalstonine	febrifuge, vermifuge, skin diseases	Wong et al., 2011
31	*Brasicca oleracea* Cruciferae	pyrrolidine,1-(1-cyclohexen-1-yl	anticancer, antidiabetic, antioxidant, antiasthmatic	Cartea et al., 2011
32	*Catharanthus roseus* Apocynaceae	catharanthine; vindoline	antioxidant, anticancer	Keerthana et al., 2021
33	*Colchicum autumnale* Colchicaceae	Colchicine; colchicoside	muscle relaxant, treatment of gout	Zárate et al., 2001
34	*Datura metel* Solanaceae	hyoscyamine; hyoscine; littorine; valtropine; fastusine; fastusinine; acetoxytropine	antiviral, anticancer, antiulcer, antifungal, antimicrobial, antispasmodic, hypoglycemic	Maheshwari et al., 2013
35	*Fissistigma latifolium* Annonaceae	liriodenine; oxoxylopine; (-)-asimilobine; (-)-anonaine; columbamine; lysicamine; (-)-remerine	antiplasmodial, antibacterial, anticancer, antifungal, antidepressant, antioxidant	Li et al., 2013
36	*Holarrhena floribunda* Apocynaceae	holarrhesine; holadienne; conessine	antimalarial, antidiarrhoeal, antidiabetic, antidysentry	Rej et al., 1976
37	*Lycoris radiate* Amaryllidaceae	lycoramine; lycorine; tazettine; ismine; homolycorine; lycoramine N-oxide; caranine; galanthamine; ungerine; hippeastrine	expectorant, emetic, anticancer	Huang et al., 2013
38	*Pachysandra procumbens* Buxaceae	pachysamine H; pachysandrine B	-	Chang et al., 2000
39	*Piper nigrum* Piperaceae	piperine; paprazine; cepharadione A; pellitorine; sylvamide	antimicrobial, aphrodisiac, carminative, stomachache, antiseptic, diuretic	Ee et al., 2008

Table 1. (Continued)

S.No.	Plant/Family	Alkaloids present	Biological activity	Ref.
40	*Solanum pseudocapsicum* Solanaceae	solacasine	anticancer, antiviral, antihypertensive, hepatoprotective, antimicrobial	Dongre et al., 2007
41	*Thalictrum foliolosum* Ranunculaceae	thalfoliolosumines A & B	diuretic, antiseptic, antibacterial, stomachache	Li et al., 2016
42	*Ailanthus altissima* Simaroubaceae	4-hydroxycanthin-6-one; 9-hydroxycanthin—6-one; 10-hydroxycanthin-6-one; 11-hydroxycanthin-6-one; canthin-6-one; 1-methoxycanthin-6-one; 4-methoxy-1-vinyl-β-carboline	antibacterial, antiviral, antioxidant, antidiarrhoel, antiinflmmatory, antipyretic, analgesic, anticancer, antiparasitic	Kim HM et al., 2016
43	*Cannabis sativa* Cannabaceae	cannabisativine; anhydrocannabisativine	antitumor, diuretic, antipyretic, analgesic, antiemetic, antiparasitic	Lozano, 2001
44	*Centaurea montana* Asteraceae	montamine; tryptamine; Moschamine; cis-moschamine; Cis-centcyamine; centcyamine	anticancer	Shoeb et al., 2006
45	*Convolvulus arvensis* Convulvulaceae	tropine; pseudotropine; tropinone; hygrine; cuscohygrine	laxative, diuretic, purgative	Todd et al., 1995
46	*Dichrostachys cinerea* Fabaceae	bisnordihydrotoxiferine	antibacterial, anthelmintic, anticonvulsant, diuretic, purgative, laxative	N'guessan-Irié et al., 2013
47	*Fritillaria thunbergii* Liliaceae	dongbeinine; dongbeirine; zhebeinine	antitussive, antiinflammatory	Kim EJ et al., 2016
48	*Hydrastatis canadensis* Ranunculaceae	berberine; canadine; hydrastine; hydrastinine; palmatine	antifungal, antiinflammatory, anticancer, hepatoprotective, antiparasitic	Jena et al., 2012

S.No.	Plant/Family	Alkaloids present	Biological activity	Ref.
49	*Macleaya cordata* Papavaraceae	cryptopine; protopine; chelidimerine; macleayine; norsanguinarine; sanguidimerine; dihydrosanguinarine	antimicrobial, anti-inflammatory, antitumor, anticancer	Liu et al. 2013
50	*Papaver rhoeas* Papaveraceae	rhoeadine; isocorydine; coptisine; stylopine; nmethylasimilobine	antimutagenic, anti-inflammatory, antidiarrhoeal, analgesic, antiulcer	Gateva et al., 2014
51	*Plumeria alba* Apocynaceae	curine; evonine; voacamine; tubocurarine chloride; syrosingopine	purgative, cardiotonic, hypotensive, diuretic, antimicrobial	Sibi et al., 2014
52	*Sanguinaria canadensis* Papavearceae	sanguinarine; chelerythrine; sanguilutine; chelilutine; allocryptopine; sanguirubine; chelirubine; protopine	antibiotic, anticancer	Furgurson et al., 2012
53	*Solanum xanthocarpum* Solanaceae	solasodine; solasonine; solasurine; solanine; solanidine	anti-inflammatory, anti-cancer, anti-allergic, antifertility	Singh and Singh, 2010
54	*Thalictrum minus* Lesser Ranunculaceae	desoxythalidastine; isothalisopavine; thaliadine; thalicthuberine	antibacterial	Mushtaq et al., 2016
55	*Ailangium lamarckii* Alaniaceae	proteometine; proteometinol; alangine; tubulosine; alangiside; isotubulosine; psychotrine; salsoline	anti-inflammatory, analgesic, diuretic, rheumatism, leprosy	Ahad et al., 2012
56	*Aspidosperma ramiflorum* Apocynaceae	ramiflorines A & B	antibacterial	Tanaka et al., 2006
57	*Cammellia sinensis* Theaceae	caffeine; theobromine	anticancer, antioxidant, antidiabetic	Suzuki and Waller, 1985
58	*Centaurea schischkinii* Asteraceae	schischkinnin; montamine	anticancer	Shoeb et al., 2005
59	*Coscinium fenestratum* Menispermaceae	palmatine; jatrorrhizine; calumbamine; berberine	antitumor, antidiabetic, anti inflammatory, hepatoprotective	Singburaudom, 2015

Table 1. (Continued)

S.No.	Plant/Family	Alkaloids present	Biological activity	Ref.
60	*Dictaminus albus* Rutaceae	haplopine; robustine; dictamine; γ-fagarine	anticancer, diuretic, abortifacient, antiseptic, vermifuge	Nam et al., 2005
61	*Fumaria capreolata* Papaveraceae	reticuline; pallidine; protopine; coclaurine; dehydrocheilanthifoline	skin disorders, rheumatism, antioxidant, Parkinson's	Zhang et al., 1993
62	*Hyoscyamus muticus* Solanaceae	hyoscyamine; scopolamine; atropine; homatropine; littorine; tropine; pseudotropine; hygrine; norhygrine	analgesic, antispasmodic, acaricidal	Oksman-Caldentey et al., 1987
63	*Mahonia manipurensis* Berberidaceae	berberine; jatrorrhizine; palmatine; oxyacanthine	anticancer, antimicrobial, antioxidant, antiproliferative, anti-inflammatory	Pfoze et al., 2014
64	*Papaver somniferum* Papaveraceae	morphine; codeine; noscapine	analgesic, antitussive	Chalise et al., 2015
65	*Plumeria rubra* Apocynaceae	plumericidine, plumerinine	antibacterial, antiinflammatory, antifungal, antipyretic, bronchitis	Ye et al., 2009
66	*Senecio cineraria* Asteraceae	senecionine; seneciphylline; jacobine; otosenine; jacoline; jaconine; usaramine; doronine; senecionine	anticancer	Tundis et al., 2007
67	*Sophora flavescens* Fabaceae	matrine; sophocarpine; sophoridine; isomatrine; mamanine; sophoranol; oxymatrine	antitumor, antiviral, antiarthritis	Lim et al., 2014
68	*Tinospora cordifolia* Menispermaceae	berberine; palmatine; magnoflorine; tinosporin; tembetarine; isocolumbin; tetrahydropalmatine	antidiabetic, antispasmodic, anticancer, antimalarial, antiallergic	Singh et al., 2003

S.No.	Plant/Family	Alkaloids present	Biological activity	Ref.
69	*Albizia gummifera* Fabaceae	budmunchiamine K; budmunchiamine G; 9-normethylbudmunchiamine K	antimalarial, antimicrobial, cytotoxic, skin disorders, stomach disorders	Rukunga and Waterman, 1996
70	*Atropa belladonna* Solanaceae	scopolamine; hyoscyamine; tropine; belladonnine; apoatropine; hydroxyhyoscyamine; norhyoscyamine	anti-inflammatory, antiasthmatic, analgesic, neuralgia, rheumatism, peptic ulcer disorders	Paul and Datta, 2011
71	*Cananga odorata* Annonaceae	sampangine; liriodenine; lysicamine; copyrine	antidiabetic, antiinflammatory, antioxidant, antimicrobial, antiulcer, antipyretic, spermatotoxic	Tan et al., 2015
72	*Cephalotaxus harringtonia* Cephlotaxaceae	harringtonine; isoharringtonine; homoharringtonine; deoxyhomoharringtonine	anticancer	Powell et al., 1972
73	*Crinum powellii* Amaryllidaceae	lycorine; 1-O-acetyl lycorine	antitumor, antiviral, antimalarial, antifungal, antibacterial, analgesic	Niño et al., 2007
74	*Duboisia myoporoides* Solanaceae	hyoscyamine; scopolamine; butropine; apoatropine; valtropine; 6 β-hydroxyhyoscyamine	cancer therapy, motion sickness	Khanam et al., 2001
75	*Gentiana olivieri* Gentianaceae	gentianine; gentianadine	antibacterial, antifungal	Mansoor et al., 2000
76	*Hyoscyamus niger* Solanaceae	hyoscyamine; scopolamine	hypotensive, vasodilator, analgesic, anticholinergic, spasmolytic, bronchodilatory	Ghorbanpour et al., 2015

Table 1. (Continued)

S.No.	Plant/Family	Alkaloids present	Biological activity	Ref.
77	*Meiogyne virgata* Annonaceae	oxoaporphines; liriodenine; lanuginosine; asimilobine; lysicamine; anonaine; remerine; nornuciferine	-	Hadiani et al., 2012
78	*Passiflora caerulea* Passifloraceae	harmol; harmine	-	Frye and Haustein, 2007
79	*Prosopis juliflora* Leguminoseae	juliflorine; julifloridine; julifloricine; juliflorinine; juliprosinene; 3-oxojuliprosine	antibacterial, antitumor, antifungal	Sathiya and Muthuchelian, 2011
80	*Sida acuta* Malvaceae	quindoline; cryptolepine	anticancer, antipyretic,	Karou et al., 2005
81	*Stemona aphylla* Stemonacaee	stemocurtisine; oxystemokerrine	anticancer	Chanmahasathien et al., 2011
82	*Toddalia asiatica* Rutaceae	nitidine; magnoflorine; 1-demethyl dicentrinone; isointegriamide; 8-acetyl norchelerythrine; 8-methoxynitidine	antimalarial, stomachache, bronchial pains	Hu et al., 2014
83	*Ancistrocladus heyneanus* Ancistrocladaceae	isoancistrocladine; yaoundamine	antimalarial, antimicrobial	Bringmann et al., 1993
84	*Bacopa monneira* Plantaginaceae	brahmine; herpestine; hydrocotyline; dihydrocotyline	antidepressant, antiparkinson, antioxidant, anti-inflammatory, analgesic, antimicrobial, cardiovascular	Al-snafi, 2013
85	*Camptotheca acuminatea* Nyssaceae	camptothecin	anticancer	Umadevi et al., 2013
86	*Chelidonium majus* Papaveraceae	sanguinarine; berberine; stylopine; protopine; chelidonine; coptisine; chelerythrine	antiulcer, antitumor, diuretic, aniviral, antibacterial, antifungal, antiasthmatic, bronchitis, jaundice	Ćirić et al., 2008

S.No.	Plant/Family	Alkaloids present	Biological activity	Ref.
87	*Croton sylvaticus* Euphorbiaceae	julocrotine; penduliflaworisin	anti-inflammatory, antipyretic	Kapingu et al., 2009
88	*Ephedra sinica* Ephedraceae	ephedrine; pseudoephedrine; methylephedrine; norephedrine;	anti-inflammatory	Gurley, 1998
89	*Glaucium flavam* Papaveraceae	glaucine; catalane; oxoglaucine; pontevedrine	anticancer, antibacterial, antifungal, analgesic, antiinflamamtory, antioxidant, hypotensive, antitussive	Arafa et al., 2016
90	*Jatropha curcus* Euphorbiaceae	5-hydroxypyrrolidin-2-one; pyrimidine-2,4-dione	antiviral, anti-HIV, antitumor, antifungal, antiulcer, wound healing activity	Dahake et al., 2013
91	*Mitragyna speciosa* Rubiaceae	corynoxine; mitragynine; speciogynine; paynantheine; corynoxine B	anti-inflammatory, analgesic, muscle relaxant	Hassan et al., 2013
92	*Prunus persica* Rosaceae	persicaside	anthelmintic, insecticidal, sedative, diuretic, expectorant, antipyretic	Rho et al., 2007
93	*Sida cordifolia* Malvaceae	ephedrine; pseudoephedrine; vasicinol; vasicinone; nmethyl-tryptophan	analgesic, antiinflammatory, hypotensive, hepatoprotective, hypoglycemic	Pattar and Jayaraj, 2012
94	*Stephania cepharantha* Menispermaceae	berberine; aromoline; cepharadione A & B; liriodenine; lysicamine; isoboldine	antiviral	Nawawi et al., 2001
95	*Tridex procumbens* Asteraceae	akuammidine; voacangine; echitamine; angustifoline; crinamidine; chitamidine	hepatoprotective, antidiabetic, antiinflammatory, hypotensive, wound healing	Ikewuchi et al., 2015
96	*Berberis microphylla* Berberidaceae	berberine; allocryptopine; calafatine; palmatine; protopine; reticuline; scoulerine; isocorydine; thalifendine; jatrorrhizine	antibacterial, antipyretic, anti-inflammatory, diarrhoea	Manosalva et al., 2016

Alkaloids have been employed by civilizations throughout history and are still in use today. These compounds are frequently mentioned in bibliographic references for their wide range of medical and therapeutic uses. As a result, alkaloids have been isolated from both marine and terrestrial sources, and humans have developed the ability to ascertain the chemical structure of numerous derivatives with varying degrees of simplicity and complexity as well as to track each compound's biological effects on living things (Munekata et al., 2021). The hundreds of medicines that are successfully used in the context of health and the treatment of various diseases have their roots in various natural sources as well as the synthesis of several alkaloids with significant therapeutic action. Alkaloids' frequently low cytotoxicity and diversity in converting into stable salt have led to a variety of medications that are simple to administer in the body without the negative effects brought on by ingesting organic and inorganic salt with poor tolerance.

Alkaloids from the natural origin as coming from the Amaryllidaceae; alkaloids derived from the Erythrina including the synthesis and pharmacological applications; curious brominated alkaloids from marine sources among several outstanding examples; mechanisms and strategies against cancer wherein certain types of alkaloid take control of important and selective form; various derivatives' technological methods have their roots in tropane; the mention of a small class of alkaloids known as oxoisoaporphines as the main therapeutic option for treating mental illnesses including depression; trabectedin's use as an alkaloid for therapeutic use in the treatment of ovarian cancer is an intriguing contribution, and finally, a thorough overview of the daphniphyllum alkaloids is provided (Sobarzo-Sánchez, 2015). This comprehensive review of the chapters, which covers the clinical treatment of diseases of various types as well as the future development of new vaccines, will help us know and understand the significance and magnitude of the alkaloids and their role in our lives. This discussion includes a variety of nitrogenated compounds, such as marine and terrestrial alkaloids. Each section will address and explore the issues they create for the future, particularly in the areas of feeding and health in general. The use of natural products to support healthcare and illness prevention has grown in popularity on a global scale. Alkaloids are significant chemical substances that provide a wealth of potential therapeutic targets (Kaur and Arora, 2015). In vitro and in vivo tests on several cancer types reveal that certain alkaloids found in natural herbs have antiviral, insecticidal, antiproliferation, antibacterial, and antimetastatic properties (Khan et al., 2016).

Conclusion

The present chapter outlines the actions of naturally occurring alkaloids, including piperine, fritillarine, berberine, matrine, and rhynchophylline, among others. The use of alkaloids as medications appears to be very promising based on the material in the literature that is summarized in this discussion, but further study and clinical trials are required before final recommendations on certain alkaloids. After that, it is hoped that this review would raise awareness of the outstanding potential that natural alkaloids possess for use in the treatment of illnesses. Biological sources of marketed and experimental plants alkaloids, the biological activity of marketed plants alkaloids, and the biological activity of experimental plants alkaloids.

References

Abat, J. K., Kumar, S. & Mohanty, A. Ethnomedicinal, Phytochemical and Ethnopharmacological Aspects of Four Medicinal Plants of Malvaceae Used in Indian Traditional Medicines: A Review. *Medicines* 4, (2017), 75.

Abdel Rahman, S. M., Abd-Ellatif, S. A., Deraz, S. F. & Khalil, A. A. Antibacterial activity of some wild medicinal plants collected from western Mediterranean coast, Egypt: Natural alternatives for infectious disease treatment. *African J. Biotechnol.* 10, (2011), 10733-10743.

Ahad, H. A., Suma Padmaja, B., Sravanthi, M., Ramyasree, P. & Kavitha, K. Phytochemical screening and anti-inflammatory actions of Alangium salviifolium root extract. *Nat. Prod. Res.* 26, (2012), 1649-53.

Ain, Q. U., Khan, H., Mubarak, M. S. & Pervaiz, A. Plant alkaloids as antiplatelet agent: Drugs of the future in the light of recent developments. *Front. Pharmacol.* 7 (2016), 292.

Al-snafi, A. E. The pharmacology of Bacopa monniera . A review. *Int. J. Pharma Sci. Res.* 4, (2013), 154-159.

Amirkia, V. & Heinrich, M. Alkaloids as drug leads-A predictive structural and biodiversity-based analysis. *Phytochem. Lett.* 10, (2014).

Aniszewski, T. *Alkaloids - Secrets of Life: Alkaloid chemistry, biological significance, applications and ecological role*. Elsevier Science (2007).

Arafa, A. M., Mohamed, M. E. S. & Eldahmy, S. I. The aerial parts of yellow horn poppy (glaucium flavum cr.) growing in Egypt: Isoquinoline alkaloids and biological activities. *J. Pharm. Sci. Res.* 8, (2016).

Aristizábal, B. & González, Á. Autoimmunity: From Bench to Bedside, Chapter 2: Innate Immune System. in *Autoimmunity: From Bench to Bedside* (2013).

Arya, A., Chahal, R., Rao, R., Rahman, M. H., Kaushik, D., Akhtar, M. F., Saleem, A., Khalifa, S., El-Seedi, H. R., Kamel, M., Albadrani, G. M., Abdel-Daim, M. M., &

Mittal, V. Acetylcholinesterase inhibitory potential of various sesquiterpene analogues for alzheimer's disease therapy. *Biomolecules* 11 (2021), 350.

Aryal, B., Niraula, P., Khadayat, K., Adhikari, B., Khatri Chhetri, D., Sapkota, B. K., Bhattarai, B. R., Aryal, N., & Parajuli, N. Antidiabetic, Antimicrobial, and Molecular Profiling of Selected Medicinal Plants. *Evid. based Complement. Altern. Med.* 2021, (2021a), 5510099.

Aryal, B., Adhikari, B., Aryal, N., Bhattarai, B. R., Khadayat, K., & Parajuli, N. LC-HRMS Profiling and Antidiabetic, Antioxidant, and Antibacterial Activities of Acacia catechu (L.f.) Willd. *Biomed Res. Int.* 2021, (2021b), 7588711.

Aryal, B., Raut, B. K., Bhattarai, S., Bhandari, S., Tandan, P., Gyawali, K., Sharma, K., Ranabhat, D., Thapa, R., Aryal, D., Ojha, A., Devkota, H. P., & Parajuli, N. Potential Therapeutic Applications of Plant-Derived Alkaloids against Inflammatory and Neurodegenerative Diseases. *Evid. based Complement. Altern. Med.* 2022, (2022), 7299778.

Ashraf, M. A. Phytochemicals as Potential Anticancer Drugs: Time to Ponder Nature's Bounty. *BioMed Research International* vol. 2020 (2020).

Atanu, F. O., Ebiloma, U. G. & Ajayi, E. I. A review of the pharmacological aspects of Solanum nigrum Linn. *Biotechnol. Mol. Biol. Rev.* 6, (2011).

Aung, T. N., Qu, Z., Kortschak, R. D. & Adelson, D. L. Understanding the effectiveness of natural compound mixtures in cancer through their molecular mode of action. *Int. J. Mol. Sci.* 18 (2017).

Awaad, A. S., Maitland, D. J. & Moneir, S. M. New alkaloids from Casimiroa edulis fruits and their pharmacological activity. *Chem. Nat. Compd.* 43, (2007).

Bacher, N., Tiefenthaler, M., Sturm, S., Stuppner, H., Ausserlechner, M. J., Kofler, R., & Konwalinka, G. Oxindole alkaloids from Uncaria tomentosa induce apoptosis in proliferating, G0/G1-arrested and bcl-2-expressing acute lymphoblastic leukaemia cells. *Br. J. Haematol.* 132, (2006), 615–622.

Badimon, L. & Vilahur, G. Thrombosis formation on atherosclerotic lesions and plaque rupture. *J. Intern. Med.* 276 (2014).

Baker, S. H. Aristotle on the Nature and Politics of Medicine. *Apeiron* 54, (2021).

Batool, S., Hasan Mahmood, M. S., Furqan, T., Karim, M. & Batool, S. In silico Analysis of Alkaloids for Therapeutic Use in Treatment of Alzheimer's Disease by Targeting Acetylcholinesterase Enzyme. *Neurol. Neurobiol.* (2021) doi:10.31487/j.nnb.2021.02.02.

Bharadwaj, K. C., Gupta, T. & Singh, R. M. Alkaloid group of Cinchona officinalis: structural, synthetic, and medicinal aspects. in *Synthesis of Medicinal Agents from Plants* (2018). doi:10.1016/b978-0-08-102071-5.00009-x.

Bhardwaj, R. & Deshmukh, R. Parkinson's Disease: An Insight into Mechanisms and Model Systems. *Int. J. Med. Res. Heal. Sci.* 7 (2018).

Borthakur, A., Bhattacharyya, S., Alrefai, W. A., Tobacman, J. K., Ramaswamy, K., & Dudeja, P. K. Platelet-activating factor-induced NF-κB activation and IL-8 production in intestinal epithelial cells are Bcl10-dependent. *Inflamm. Bowel Dis.* 16, (2010), 593-603.

Bothara, S. B., Marya, B. H. & Saluja, A. K. Antiarthritic activity of root extracts of Cocculus hirsutus. *Int. J. Pharm. Pharm. Sci.* 3, (2011).

Bribi, N., Algieri, F., Rodriguez-Nogales, A., Garrido-Mesa, J., Vezza, T., Maiza, F., Utrilla, M. P., Rodriguez-Cabezas, M. E., & Galvez, J. Antinociceptive and anti-inflammatory effects of total alkaloid extract from fumaria capreolata. *Evid. based Complement. Altern. Med.* 2015, (2015), 736895.

Bringmann, G., Kinzinger, L., Ortmann, T. & Souza, N. J. D. Isoancistrocladine from ancistrocladus heyneanus: The first naturally occurring n-unsubstituted cis-configurated naphthyltetrahydroisoquinoline alkaloid. *Phytochemistry* 35, (1993).

Cantrell, C. L., Abate, L., Fronczek, F. R., Franzblau, S. G., Quijano, L., & Fischer, N. H. Antimycobacterial eudesmanolides from Inula helenium and Rudbeckia subtomentosa. *Planta Med.* 65, (1999), 351–355.

Carroll, H. P., Paunović, V. & Gadina, M. Signalling, inflammation and arthritis: Crossed signals: The role of interleukin-15 and -18 in autoimmunity. *Rheumatology* vol. 47 (2008).

Cartea, M. E., Francisco, M., Soengas, P. & Velasco, P. Phenolic compounds in Brassica vegetables. *Molecules* vol. 16 (2011).

Casciaro, B., Mangiardi, L., Cappiello, F., Romeo, I., Loffredo, M. R., Iazzetti, A., Calcaterra, A., Goggiamani, A., Ghirga, F., Mangoni, M. L., Botta, B., & Quaglio, D. Naturally-occurring alkaloids of plant origin as potential antimicrobials against antibiotic-resistant infections. *Molecules* vol. 25 (2020).

Chalise, U. The Poppy Plant: Phytochemistry & Pharmacology. *Indo Glob. J. Pharm. Sci.* 05, (2015).

Chang, L. C., Bhat, K. P. L., Fong, H. H. S., Pezzuto, J. M. & Kinghorn, A. D. Novel bioactive steroidal alkaloids from Pachysandra procumbens. *Tetrahedron* 56, (2000).

Changxing, L., Galani, S., Hassan, F. U., Rashid, Z., Naveed, M., Fang, D., Ashraf, A., Qi, W., Arif, A., Saeed, M., Chishti, A. A., & Jianhua, L. Biotechnological approaches to the production of plant-derived promising anticancer agents: An update and overview. *Biomed. Pharmacother.* 132 (2020), 110918.

Chanmahasathien, W., Ampasavate, C., Greger, H. & Limtrakul, P. Stemona alkaloids, from traditional Thai medicine, increase chemosensitivity via P-glycoprotein-mediated multidrug resistance. *Phytomedicine* 18, (2011).

Cheenpracha, S., Jitonnom, J., Komek, M., Ritthiwigrom, T. & Laphookhieo, S. Acetylcholinesterase inhibitory activity and molecular docking study of steroidal alkaloids from Holarrhena pubescens barks. *Steroids* 108, (2016).

Chen, P. Y., Wu, M. S., Chen, C. Y., Bair, M. J., Chou, C. K., Lin, J. T., Liou, J. M., & Taiwan Gastrointestinal Disease and Helicobacter Consortium. Systematic review with meta-analysis: the efficacy of levofloxacin triple therapy as the first- or second-line treatments of Helicobacter pylori infection. *Aliment. Pharmacol. Ther.* 44 (2016), 427–437.

Chothani, D. L. & Vaghasiya, H. U. A phyto-pharmacological overview on physalis minima linn. *Indian Journal of Natural Products and Resources* vol. 3 (2012).

Ćirić, A., Vinterhalter, B., Šavikin-Fodulović, K., Soković, M. & Vinterhalter, D. Chemical analysis and antimicrobial activity of methanol extracts of celandine (Chelidonium majus L.) plants growing in nature and cultured in vitro. *Arch. Biol. Sci.* 60, (2008).

Cronin, D. A. Principles and applications of gas chromatography in food analysis. *Trends Food Sci. Technol.* 1, (1990).

D'Orazio, N., Gammone, M. A., Gemello, E., De Girolamo, M., Cusenza, S., & Riccioni, G. Marine bioactives: Pharmacological properties and potential applications against inflammatory diseases. *Mar. Drugs* 10 (2012), 812–833.

Dahake, R., Roy, S., Patil, D., Rajopadhye, S., Chowdhary, A., Deshmukh, R. A. Potential Anti-HIV activity of jatropha curcas Linn. Leaf extracts. *J. Antivirals Antiretrovir.* 5, (2013), 160-165.

Dall'Acqua, S. Plant-derived acetylcholinesterase inhibitory alkaloids for the treatment of Alzheimer's disease. *Bot. Targets Ther.* 3 (2013), 19-28.

Dauer, W. & Przedborski, S. Parkinson's disease: Mechanisms and models. *Neuron* vol. 39 (2003).

Davis, B. J. & Roughley, S. D. Fragment-Based Lead Discovery. in *Ann. Rep. Med. Chem.* 50 (2017).

Debnath, B., Singha, W. S., Das, M., Goswami, S., Singh, M. K., Maiti, D., Manna, K. Role of plant alkaloids on human health: A review of biological activities. *Materials Today Chemistry* vol. 9 (2018), 56-72.

Dénès, F. (2018). Heteroatom-centred radicals for the synthesis of heterocyclic compounds. In Y. Landais (Ed.), *Free-Radical Synthesis and Functionalization of Heterocycles* (Vol. 54, pp. 151–230). Springer International Publishing. https://doi.org/10.1007/7081_2018_19.

Dholwani, K., Saluja, A., Gupta, A. & Shah, D. A review on plant-derived natural products and their analogs with anti-tumor activity. *Indian J. Pharmacol.* 40 (2008).

Dongre, S., Badami, S., Ashok, G., Ravi, S. & Kottani, R. In vitro cytotoxic properties of O-methylsolanocapsine isolated from Solanum pseudocapsicum leaves. *Indian J. Pharmacol.* 39, (2007).

Durairajan, S. S., Liu, L. F., Lu, J. H., Chen, L. L., Yuan, Q., Chung, S. K., Huang, L., Li, X. S., Huang, J. D., & Li, M. Berberine ameliorates β-amyloid pathology, gliosis, and cognitive impairment in an Alzheimer's disease transgenic mouse model. *Neurobiol. Aging* 33, (2012), 2903–2919.

Ee, G. C. L., Lim, S. K., Lim, C. M. & Dzulkefly, K. Alkaloids and carboxylic acids from Piper nigrum. *Asian J. Chem.* 20, (2008).

Emwas, A. H., Roy, R., McKay, R. T., Tenori, L., Saccenti, E., Gowda, G., Raftery, D., Alahmari, F., Jaremko, L., Jaremko, M., & Wishart, D. S. Nmr spectroscopy for metabolomics research. *Metabolites* 9 (2019), 123.

Erasto, P., Mbwambo, Z. H., Nondo, R. S. O., Lall, N. & Lubschagne, A. Antimycobacterial, antioxidant activity and toxicity of extracts from the roots of Rauvolfia vomitoria and R. caffra Rauvolfia vomitoria ve R. Caffra kök ekstraktlarının antimikobakteriyel, antioksidan etkinliği ve toksisitesi. *Orig. Artic. Spat. DD* 1, (2011).

Feng, T., Duan, K., He, S., Wu, B., Zheng, Y., Ai, H., Li, Z., He, J., Zuo, J., & Liu, J. Ophiorrhines A and B, Two Immunosuppressive Monoterpenoid Indole Alkaloids from Ophiorrhiza japonica. *Org. Lett.* 20, (2018), 7926-7928.

Fielding, B. C., Filho, C. da S. M. B., Ismail, N. S. M. & de Sousa, D. P. Alkaloids: Therapeutic potential against human coronaviruses. *Molecules* 25, (2020).

Frye, A. & Haustein, C. Extraction, Identification, and Quantification of Harmala Alkaloids in Three Species of Passiflora. *Am. J. Undergrad. Res.* 6, (2007).

Furgurson, J., Cubbage, F., Sills, E. & Bates, P. Bloodroot (Sanguinaria canadensis L.) Extent and Sustainability in Western North Carolina. *Open J. For.* 02, (2012).

Furman, D., Furman, D., Campisi, J., Verdin, E., Carrera-Bastos, P., Targ, S., Franceschi, C., Ferrucci, L., Gilroy, D. W., Fasano, A., Miller, G. W., Miller, A. H., Mantovani, A., Weyand, C. M., Barzilai, N., Goronzy, J. J., Rando, T. A., Effros, R. B., Lucia, A., Kleinstreuer, N., & Slavich, G. M. Chronic inflammation in the etiology of disease across the life span. *Nat. Med.* 25, (2019), 1822–1832.

Gangwar, A. K. & Ghosh, A. K. Medicinal uses and Pharmacological activity of Adhatoda vasica. *Int. J. Herb. Med.* 2, (2014).

Gao, H., Wang, W., Chu, W., Liu, K., Liu, Y., Liu, X., Yao, H., & Gao, Q. Paniculatumoside g, a new c21 steroidal glycoside from cynanchum paniculatum. *Rev. Bras. Farmacogn.* 27, (2017), 54-58.

Gao, L. N., Feng, Q. S., Zhang, X. F., Wang, Q. S. & Cui, Y. L. Tetrandrine suppresses articular inflammatory response by inhibiting pro-inflammatory factors via NF-κB inactivation. *J. Orthop. Res.* 34, (2016).

Gateva, S., Jovtchev, G., Stankov, A. & Gregan, F. Antigenotoxic capacity of Papaver rhoeas L. Extract. *Int. J. Pharm. Pharm. Sci.* 6, (2014).

Ghorbanpour, M., Hatami, M. & Hatami, M. Activating antioxidant enzymes, hyoscyamine and scopolamine biosynthesis of Hyoscyamus Niger L. Plants with nano-sized titanium dioxide and bulk Application. *Acta Agric. Slov.* 105, (2015).

Gunaydin, C. & Bilge, S. S. Effects of nonsteroidal anti-inflammatory drugs at the molecular level. *Eurasian Journal of Medicine* vol. 50 (2018).

Guo, H., Callaway, J. B. & Ting, J. P. Y. Inflammasomes: Mechanism of action, role in disease, and therapeutics. *Nature Medicine* vol. 21 (2015).

Gurley, B. J. Ephedrine-type alkaloid content of nutritional supplements containing Ephedra sinica (Ma-huang) as determined by high performance liquid chromatography. *J. Pharm. Sci.* 87, (1998).

Hadiani, N., Alias, A. & Puteh, C. Alkaloids and Anthraquinones from Malaysian Flora. in *Phytochemicals - A Global Perspective of Their Role in Nutrition and Health* (2012). doi:10.5772/26753.

Hall, R. & Mazer, C. D. Antiplatelet drugs: A review of their pharmacology and management in the perioperative period. *Anesthesia and Analgesia* vol. 112 (2011).

Hallmann-Mikołajczak, A. Ebers Papyrus. The book of medical knowledge of the 16th century B. C. Egyptians. *Arch. Hist. Filoz. Med.* 67, (2004).

Harvey, A. L. Natural products in drug discovery. *Drug Discovery Today* vol. 13 (2008).

Hassan, Z., Muzaimi, M., Navaratnam, V., Yusoff, N. H., Suhaimi, F. W., Vadivelu, R., Vicknasingam, B. K., Amato, D., von Hörsten, S., Ismail, N. I., Jayabalan, N., Hazim, A. I., Mansor, S. M., & Müller, C. P. From Kratom to mitragynine and its derivatives: Physiological and behavioural effects related to use, abuse, and addiction. *Neurosci. Biobehav. Rev.* 37 (2013), 138–151.

He, F. Q., Qiu, B. Y., Zhang, X. H., Li, T. K., Xie, Q., Cui, D. J., Huang, X. L., & Gan, H. T. Tetrandrine attenuates spatial memory impairment and hippocampal neuroinflammation via inhibiting NF-κB activation in a rat model of Alzheimer's disease induced by amyloid-β(1-42). *Brain Res.* 1384, (2011), 89–96.

Heinrich, M., Mah, J. & Amirkia, V. Alkaloids used as medicines: Structural phytochemistry meets biodiversity—An update and forward look. *Molecules* vol. 26 (2021).

Hou, J. P. The development of Chinese herbal medicine and the Pen-ts'ao. *Comp. Med. East West* 5, (1977).

Hu, J., Shi, X., Chen, J., Mao, X., Zhu, L., Yu, L., Shi, J. Alkaloids from Toddalia asiatica and their cytotoxic, antimicrobial and antifungal activities. *Food Chem.* 148, (2014), 437-444.

Huang, J., Liu, Y., Chen, J. X., Lu, X. Y., Zhu, W. J., Qin, L., Xun, Z. X., Zheng, Q. Y., Li, E. M., Sun, N., Xu, C., & Chen, H. Y. Harmine is an effective therapeutic small molecule for the treatment of cardiac hypertrophy. *Acta Pharmacol. Sin.* 43, (2022), 50–63.

Huang, S. D., Zhang, Y., He, H. P., Li, S. F., Tang, G. H., Chen, D. Z., Cao, M. M., DI, Y. T., & Hao, X. J. A new amaryllidaceae alkaloid from the bulbs of lycoris radiata. *Chin. J. Nat. Med.* 11, (2013), 406-410.

Hussain, G., Rasul, A., Anwar, H., Aziz, N., Razzaq, A., Wei, W., Ali, M., Li, J., & Li, X. Role of plant derived alkaloids and their mechanism in neurodegenerative disorders. *Int. J. Biol. Sci.* 14, (2018), 341–357.

Ikewuchi, C. C., Ikewuchi, J. C. & Ifeanacho, M. O. Phytochemical Composition of Tridax procumbens Linn Leaves: Potential as a Functional Food. *Food Nutr. Sci.* 06, (2015), 992-1004.

Iqbal, J., Abbasi, B. A., Mahmood, T., Kanwal, S., Ali, B., Shah, S. A., & Khalil, A. T. Plant-derived anticancer agents: A green anticancer approach. *Asian Pac. J. Trop. Biomed.* 7 (2017), 1129-1150.

Jena, J., Ranjan, R., Ranjan, P. & Sarangi, M. K. A Study on Natural Anticancer Plants. *Int. J. Pharm. Chem. Sci.* 1, (2012).

Jiang, C., Ma, Q., Wang, S., Shen, Y., Qin, A., Fan, S., & Jie, Z. Oxymatrine Attenuates Osteoclastogenesis via Modulation of ROS-Mediated SREBP2 Signaling and Counteracts Ovariectomy-Induced Osteoporosis. *Front. Cell Dev. Biol.* 9, (2021), 684007.

Jiang, X. L., Wang, L., Wang, E. J., Zhang, G. L., Chen, B., Wang, M. K., & Li, F. Flavonoid glycosides and alkaloids from the embryos of Nelumbo nucifera seeds and their antioxidant activity. *Fitoterapia* 125, (2018), 184-190.

Kapingu, M., Mbwambo, Z., Moshi, M. & Magadula, J. Brine shrimp lethality of a glutarimide alkaloid from *Croton sylvaticus* Hochst. *East Cent. African J. Pharm. Sci.* 8, (2006).

Karou, D., Savadogo, A., Canini, A., Yameogo, S., Montesano, C., Simpore, J., Colizzi, V., & Traore, A. S. Antibacterial activity of alkaloids from Sida acuta. *African J. Biotechnol.* 4, (2005), 1452-1457.

Kaur, R. & Arora, S. Alkaloids-Important Therapeutic Secondary Metabolites of Plant Origin. *J. Crit. Rev.* 2, (2015).

Keerthana, S., Mahalakshmi, S., Kavitha, S., Ramesh Babu, N. G., Sivaraj, C., & Arumugam, P. Antioxidant and Anticancer Activities of Roots of Catharanthus roseus (L.) G.Don. *Res. J. Chem. Environ.* 25, (2021), 158-165.

Khan, H., Mubarak, M. S. & Amin, S. Antifungal Potential of Alkaloids As An Emerging Therapeutic Target. *Curr. Drug Targets* 18, (2016).

Khanam, N., Khoo, C., Close, R. & Khan, A. G. Tropane alkaloid production by shoot culture of Duboisia myoporoides R. Br. *Phytochemistry* 56, (2001).

Kim, E. J., Yoon, Y. P., Woo, K. W., Kim, J. H., Min, S. Y., Lee, H. J., Lee, S. K., Hong, J. H., Lee, K. R., & Lee, C. J. Verticine, ebeiedine and suchengbeisine isolated from the bulbs of Fritillaria thunbergii Miq. inhibited the gene expression and production of MUC5AC mucin from human airway epithelial cells. *Phytomedicine* 23, (2016), 95-104.

Kim, H. M., Lee, J. S., Sezirahiga, J., Kwon, J., Jeong, M., Lee, D., Choi, J. H., & Jang, D. S. A new canthinone-type alkaloid isolated from Ailanthus altissima Swingle. *Molecules* 21, (2016), 642.

Kim, M., Cho, K. H., Shin, M. S., Lee, J. M., Cho, H. S., Kim, C. J., Shin, D. H., & Yang, H. J. Berberine prevents nigrostriatal dopaminergic neuronal loss and suppresses hippocampal apoptosis in mice with Parkinson's disease. *Int. J. Mol. Med.* 33, (2014), 870–878.

Koch, P., Gehringer, M. & Laufer, S. A. Inhibitors of c-Jun N-terminal kinases: An update. *J. Med. Chem.* 58, (2015).

Krikorian, A. D. A Dictionary of Natural Products: Terms in the Field of Pharmacognosy, Relating to Natural Medicinal and Pharmaceutical Materials and the Plants, Animals, and Minerals from Which They are Derived. George Macdonald Hocking . *Q. Rev. Biol.* 73, (1998).

Krishnaiah, D., Sarbatly, R. & Nithyanandam, R. A review of the antioxidant potential of medicinal plant species. *Food and Bioproducts Processing* vol. 89 (2011).

Kurhekar, J. V. Ancient and modern practices in phytomedicine. in *Preparation of Phytopharmaceuticals for the Management of Disorders: The Development of Nutraceuticals and Traditional Medicine.* Academic Press (2020).

Lambole, V. B., Murti, K., Kumar, U., Bhatt Sandipkumar, P. & Gajera, V. Phytopharmacological properties of Aegle marmelos as a potential medicinal tree: An overview. *International Journal of Pharmaceutical Sciences Review and Research* vol. 5 (2010).

Levi, R., Capurro, N. & Lee, C. H. Pharmacological characterization of cardiac histamine receptors: Sensitivity to H1- and H2-receptor agonists and antagonists. *Eur. J. Pharmacol.* 30, (1975).

Li, D. H., Guo, J., Bin, W., Zhao, N., Wang, K. B., Li, J. Y., Li, Z. L., & Hua, H. M. Two new benzylisoquinoline alkaloids from Thalictrum foliolosum and their antioxidant and in vitro antiproliferative properties. *Arch. Pharm. Res.* 39, (2016), 871–877.

Li, H. T., Wu, H. M., Chen, H. L., Liu, C. M. & Chen, C. Y. The pharmacological activities of (-)-anonaine. *Molecules* vol. 18 (2013).

Liang, G., Zhou, H., Wang, Y., Gurley, E. C., Feng, B., Chen, L., Xiao, J., Yang, S., & Li, X. Inhibition of LPS-induced production of inflammatory factors in the macrophages by mono-carbonyl analogues of curcumin. *J. Cell. Mol. Med.* 13, (2009), 3370–3379.

Lim, S. J., Jeong, D. Y., Choi, G. H., Park, B. J. & Kim, J. H. Quantitative Analysis of Matrine and Oxymatrine in Sophora flavescens Extract and Its Biopesticides by UPLC. *J. Agric. Chem. Environ.* 03, (2014).

Lisko, J. G., Lee, G. E., Kimbrell, J. B., Rybak, M. E., Valentin-Blasini, L., & Watson, C. H. Caffeine Concentrations in Coffee, Tea, Chocolate, and Energy Drink Flavored E-liquids. *Nicotine Tob. Res.* 19, (2017), 484–492.

Liu, M., Lin, Y. ling, Chen, X. R., Liao, C. C. & Poo, W. K. In vitro assessment of Macleaya cordata crude extract bioactivity and anticancer properties in normal and cancerous human lung cells. *Exp. Toxicol. Pathol.* 65, (2013).

Liu, Q., He, H., Mai, L., Yang, S., Fan, W., & Huang, F. Peripherally Acting Opioids in Orofacial Pain. *Front. Neurosci.* 15 (2021), 665445.

Liu, Y., Xu, Y., Ji, W., Li, X., Sun, B., Gao, Q., & Su, C. Anti-tumor activities of matrine and oxymatrine: Literature review. *Tumor Biol.* 35 (2014).

Liu, Y., Zhang, Y., Zheng, X., Fang, T., Yang, X., Luo, X., Guo, A., Newell, K. A., Huang, X. F., & Yu, Y. Galantamine improves cognition, hippocampal inflammation, and synaptic plasticity impairments induced by lipopolysaccharide in mice. *J. Neuroinflammation* 15, (2018), 112.

Lobo, V., Patil, A., Phatak, A. & Chandra, N. Free radicals, antioxidants and functional foods: Impact on human health. *Pharmacognosy Reviews* vol. 4 (2010).

Lou, T., Zhang, Z., Xi, Z., Liu, K., Li, L., Liu, B., & Huang, F. Berberine inhibits inflammatory response and ameliorates insulin resistance in hepatocytes. *Inflammation* 34, (2011), 659–667.

Lozano, I. The therapeutic use of Cannabis sativa (L.) in Arabic medicine. *J. Cannabis Ther.* 1, (2001).

Lu, J. J., Bao, J. L., Chen, X. P., Huang, M. & Wang, Y. T. Alkaloids isolated from natural herbs as the anticancer agents. *Evid. based Complement. Alternat. Med.* 2012 (2012).

Ma, Z. Z., Hano, Y., Nomura, T. & Chen, Y. J. Alkaloids and phenylpropanoids from Peganum nigellastrum. *Phytochemistry* 53, (2000).

Madan, A. K. & Dureja, H. Prediction of pharmacokinetic parameters. *Methods Mol. Biol.* 929, (2012).

Maheshwari, N. O., Khan, A. & Chopade, B. A. Rediscovering the medicinal properties of Datura sp .: A review. *J. Med. Plants Res.* 7, (2013).

Majnooni, M. B., Fakhri, S., Bahrami, G., Naseri, M., Farzaei, M. H., & Echeverría, J. Alkaloids as Potential Phytochemicals against SARS-CoV-2: Approaches to the Associated Pivotal Mechanisms. *Evid. based Complement. Alternat. Med.* 2021 (2021), 6632623.

Manosalva, L., Mutis, A., Urzúa, A., Fajardo, V. & Quiroz, A. Antibacterial activity of alkaloid fractions from berberis microphylla G. Forst and study of synergism with ampicillin and cephalothin. *Molecules* 21, (2016).

Mansoor, A., Zaidi, M. I. & Malghani, M. A. K. Isolation of bioactive alkaloids from Gentiana olivieri and its non-toxic effect. *Pakistan J. Bot.* 32, (2000).

Marya, B. H. & Bothara, S. B. Ethnopharmacological properties of cocculus hirsutus (L.) diels- a review. *International Journal of Pharmaceutical Sciences Review and Research* vol. 7 (2011).

Matsuura, H. N. & Fett-Neto, A. G. Plant Alkaloids: Main Features, Toxicity, and Mechanisms of Action. in *Plant Toxins* (2015). doi:10.1007/978-94-007-6728-7_2-1.

Mei, L., Deng, Y., Li, F., Guo, D. le & Lu, C. yu. [Study on the alkaloids constituents from Voacanga africana]. *Zhong Yao Cai* 35, (2012).

Melhaoui, A., Mallear, M., Jossang, A. & Bodo, B. Antibiotic and antifungal pyrrolidine alkaloids, from arisarum vulgare. *Nat. Prod. Lett.* 2, (1993).

Monte, J., Abreu, A. C., Borges, A., Simões, L. C. & Simões, M. Antimicrobial activity of selected phytochemicals against Escherichia coli and Staphylococcus aureus and their biofilms. *Pathogens* 3, (2014).

Müller-Schüssele, S. J., Schwarzländer, M. & Meyer, A. J. Live monitoring of plant redox and energy physiology with genetically encoded biosensors. *Plant Physiology* vol. 186 (2021).

Munekata, P., Pateiro, M., Conte-Junior, C. A., Domínguez, R., Nawaz, A., Walayat, N., Movilla Fierro, E., & Lorenzo, J. M. Marine alkaloids: Compounds with in vivo activity and chemical synthesis. *Mar. Drugs* 19 (2021), 374.

Mushtaq, S., Rather, M. A., Qazi, P. H., Aga, M. A., Shah, A. M., Shah, A., & Ali, M. N. Isolation and characterization of three benzylisoquinoline alkaloids from Thalictrum minus L. and their antibacterial activity against bovine mastitis. *J. Ethnopharmacol.* 193, (2016), 221–226.

Mythili, S., Gajalakshmi, S., Sathiavelu, A. & Sridharan, T. B. Pharmacological Activities of Cassytha Filiformis: A Review. *Asian J. Plant Sci. Res.* 1, (2011).

N'guessan-Irié, A. G., Dade, J., Champy, P., Siransy-Kouakou, N. G. & Leblais, V. Isolation and Spasmolytic Evaluation of New Alkaloids from Dichrostachys cinerea (L.) Wight et Arn. (Fabaceae). *Pharmacol. Pharm.* 04, (2013), 684-688.

Nam, K. W., Je, K. H., Shin, Y. J., Sam, S. K. & Mar, W. Inhibitory effects of furoquinoline alkaloids from Melicope confusa and Dictamnus albus against human phosphodiesterase 5 (hPDE5A) in vitro. *Arch. Pharm. Res.* 28, (2005)

Nawawi, A., Nakamura, N., Meselhy, M. R., Hattori, M., Kurokawa, M., Shiraki, K., Kashiwaba, N., & Ono, M. In vivo antiviral activity of Stephania cepharantha against herpes simplex virus type-1. *Phyther. Res.* 15, (2001), 497–500.

Ng, Y. P., Or, T. C. T. & Ip, N. Y. Plant alkaloids as drug leads for Alzheimer's disease. *Neurochem. Int.* 89 (2015).

Ngo, Q. M., Tran, P. T., Tran, M. H., Kim, J. A., Rho, S. S., Lim, C. H., Kim, J. C., Woo, M. H., Choi, J. S., Lee, J. H., & Min, B. S. Alkaloids from Piper nigrum Exhibit Antiinflammatory Activity via Activating the Nrf2/HO-1 Pathway. *Phyther. Res.* 31, (2017), 663–670.

Nigam, M. Phytomedicine: Scope and current highlights. in *Preparation of Phytopharmaceuticals for the Management of Disorders: The Development of Nutraceuticals and Traditional Medicine.* Elsevier (2020), pp. 39–54.

Niño, J., Hincapié, G. M., Correa, Y. M. & Mosquera, O. M. Alkaloids of Crinum x powellii "Album" (Amaryllidaceae) and their Topoisomerase Inhibitory Activity. *Z. Naturforsch. C* 62, (2007).

Novelli, S., Lorena, C. & Antonella, C. Identification of Alkaloid's Profile in Ficus benjamina L. Extracts with Higher Antioxidant Power. *Am. J. Plant Sci.* 05, (2014).

Obaid Aldulaimi, A. K., Azziz, S. S. A., Bakri, Y. M., Nafiah, M. A., Aowda, S. A., Awang, K., & Litaudon, M. Two New isoquinoline alkaloids from the bark of Alphonsea cylindrica King and their antioxidant activity. *Phytochem. Lett.* 29, (2019), 110-114.

Oksman-Caldentey, K. M., Vuorela, H., Strauss, A. & Hiltunen, R. Variation in the tropane alkaloid content of Hyoscyamus muticus plants and cell culture clones. *Planta Med.* 53, (1987).

Panahi, N., Mahmoudian, M., Mortazavi, P. & Hashjin, G. S. Effects of berberine on β-secretase activity in a rabbit model of Alzheimer's disease. *Arch. Med. Sci.* 9, (2013).

Patel, K., Gadewar, M., Tripathi, R., Prasad, S. K. & Patel, D. K. A review on medicinal importance, pharmacological activity and bioanalytical aspects of beta-carboline alkaloid 'Harmine'. *Asian Pac. J. Trop. Biomed.* 2, (2012).

Patil, S. P. Recently isolated lycodine-type Lycopodium alkaloids and their total synthesis: a review. *Futur. J. Pharm. Sci.* 6, (2020).

Patnala, S. & Kanfer, I. HPLC analysis of mesembrine-type alkaloids in Sceletium plant material used as an african traditional medicine. *J. Pharm. Pharm. Sci.* 13, (2010).

Pattar, P. V. & Jayaraj, M. Pharmacognostic and phytochemical investigation of sida cordifolia l.- A threatened medicinal herb. *Int. J. Pharm. Pharm. Sci.* 4, (2012).

Patwardhan, B. Ethnopharmacology and drug discovery. *Journal of Ethnopharmacology* vol. 100 (2005).

Paul, R. & Datta K. A. An updated overview on atropa belladonna L. *Int. Res. J. Pharm.* 2, (2011).

Petrovska, B. B. Historical review of medicinal plants' usage. *Pharmacognosy Reviews* vol. 6 (2012).

Pfoze, N. L., Myrboh, B., Kumar, Y. & Rohman, M. R. Isolation of Protoberberine Alkaloids from Stem Bark of Mahonia manipurensis Takeda Using RP-HPLC. *J. Med. Plants Stud.* 2, (2014).

Poddar, M. K., Chakraborty, A. & Banerjee, S. Neurodegeneration: Diagnosis, Prevention, and Therapy. in *Oxidoreductase* (2021). doi:10.5772/intechopen.94950.

Polis, B. & Samson, A. O. Arginase as a Potential Target in the Treatment of Alzheimer's Disease. *Adv. Alzheimer's Dis.* 07, (2018).

Powell, R. G., Weisleder, D. & Smith, C. R. Antitumor alkaloids from Cephalotaxus harringtonia: Structure and activity. *J. Pharm. Sci.* 61, (1972).

Puri, S. K., Habbu, P. V., Kulkarni, P. V. & Kulkarni, V. H. Nitrogen containing secondary metabolites from endophytes of medicinal plants and their biological/pharmacological activities-a review. *Sys. Rev. Pharm.* 9 (2018), 22-30.

Qiu, S., Sun, H., Zhang, A. H., Xu, H. Y., Yan, G. L., Han, Y., & Wang, X. J. Natural alkaloids: Basic aspects, biological roles, and future perspectives. *Chin. J. Nat. Med.* 12, (2014), 401–406.

Rajput, A., Sharma, R. & Bharti, R. Pharmacological activities and toxicities of alkaloids on human health. in *Materials Today: Proceedings* vol. 48 (2021).

Raza, C., Anjum, R. & Shakeel, N. ul A. Parkinson's disease: Mechanisms, translational models and management strategies. *Life Sci*. 226 77–90 (2019).

Rej, R. N., Ghosh, P. C. & Banerji, J. Holacetine, a new steroid alkaloid from Holarrhena antidysenterica. *Phytochemistry* 15, (1976).

Rho, J. R., Ha, Y. A., Cui, M. X., Lim, J. A., & Chai, K. Y. Isolation and characterization of a new alkaloid from the seed of Prunus persica L. and its anti-inflammatory activity. *Bull. Korean Chem. Soc.* 28, (2007), 1289-1293.

Richardson, W. M., Sodhi, C. P., Russo, A., Siggers, R. H., Afrazi, A., Gribar, S. C., Neal, M. D., Dai, S., Prindle, T., Jr, Branca, M., Ma, C., Ozolek, J., & Hackam, D. J. Nucleotide-binding oligomerization domain-2 inhibits toll-like receptor-4 signaling in the intestinal epithelium. *Gastroenterology* 139, (2010), 904–917.

Roy, A. & Chowdhury, G. Pharmacological Activities of Indian Heliotrope (Heliotropium Indicum L.): A Review. *J. Pharmacogn. Phytochem.* 4, (2015).

Roy, N. S., Choi, I. Y., Um, T., Jeon, M. J., Kim, B. Y., Kim, Y. D., Yu, J. K., Kim, S., & Kim, N. S. Gene expression and isoform identification of pacbio full-length cdna sequences for berberine biosynthesis in berberis koreana. *Plants* 10, (2021), 1314.

Rubio-Pina, J. & Vazquez-Flota, F. Pharmaceutical Applications of the Benzylisoquinoline Alkaloids from Argemone mexicana L. *Curr. Top. Med. Chem.* 13, (2013).

Rukunga, G. M. & Waterman, P. G. New macrocyclic spermine (budmunchiamine) alkaloids from Albizia gummifera: With some observations on the structure-activity relationships of the budmunchiamines. *J. Nat. Prod.* 59, (1996).

Runtao, G., Guo, D., Jiangbo, Y., Xu, W. & Shusen, Y. Oxymatrine, the main alkaloid component of Sophora roots, protects heart against arrhythmias in rats. *Planta Med.* 77, (2011).

Sahu, N. P. & Mahato, S. B. Anti-inflammatory triterpene saponins of Pithecellobium dulce: Characterization of an echinocystic acid bisdesmoside. *Phytochemistry* 37, (1994).

Sainsbury, M. & Webb, B. Minor alkaloids of bleekeria vitiensis. *Phytochemistry* 11, (1972).

Satake, H., Ono, E. & Murata, J. Recent advances in the metabolic engineering of lignan biosynthesis pathways for the production of transgenic plant-based foods and supplements. in *Journal of Agricultural and Food Chemistry* vol. 61 (2013).

Sathiya, M. & Muthuchelian, K. Anti-tumor potential of total alkaloid extract of Prosopisjuliflora DC. leaves against Molt-4 cells in vitro. *African J. Biotechnol.* 10, (2011).

Scheller, J., Chalaris, A., Schmidt-Arras, D. & Rose-John, S. The pro- and anti-inflammatory properties of the cytokine interleukin-6. *Biochim. Biophys. Acta Mol. Cell Res.* 1813 (2011).

Schmeller, T. & Wink, M. Utilization of Alkaloids in Modern Medicine. in *Alkaloids* (1998). doi:10.1007/978-1-4757-2905-4_18.

Sharma, P., Jha, A. B., Dubey, R. S. & Pessarakli, M. Reactive Oxygen Species, Oxidative Damage, and Antioxidative Defense Mechanism in Plants under Stressful Conditions. *J. Bot.* 2012, (2012).

Shawky, E., El Sohafy, S. M., de Andrade, J. P. & de Souza Borges, W. Profiling of acetylcholinesterase inhibitory alkaloids from some Crinum, Habranthus and Zephyranthes species by GC–MS combined with multivariate analyses and in silico studies. *Nat. Prod. Res.* 35, (2021).

Shoeb, M., Celik, S., Jaspars, M., Kumarasamy, Y., MacManus, S. M., Nahar, L., Thoo-Lin, P. K., & Sarker, S. D. Isolation, structure elucidation and bioactivity of schischkiniin, a unique indole alkaloid from the seeds of Centaurea schischkinii. *Tetrahedron* 61, (2005).

Shoeb, M., Macmanus, S. M., Jaspars, M., Trevidu, J., Nahar, L., Kong-Thoo-Lin, P., & Sarker, S. D. Montamine, a unique dimeric indole alkaloid, from the seeds of Centaurea montana (Asteraceae), and its in vitro cytotoxic activity against the CaCo2 colon cancer cells. *Tetrahedron* 62, (2006), 11172-11177.

Sibi, G., Venkategowda, A. & Gowda, L. Isolation and Characterization of Antimicrobial Alkaloids from Plumeria alba Flowers against Food Borne Pathogens. *Am. J. Life Sci.* 2, (2014), 1-6.

Sichaem, J., Tip-Pyang, S. & Siripong, P. Chemical constituents from the root bark of calophyllum inophyllum. *Nat. Prod. Commun.* 13, (2018).

Simmonds, R. E. & Foxwell, B. M. Signalling, inflammation and arthritis: NF-κB and its relevance to arthritis and inflammation. *Rheumatology* 47, (2008).

Singburaudom, N. The alkaloid Berberine isolated from Coscinium fenestratum is an inhibitor of phytopathogenic fungi. *J. Biopestic.* 8, (2015).

Singh, O. M. & Singh, T. P. Phytochemistry of Solanum xanthocarpum: An amazing traditional healer. *J. Sci. Ind. Res. (India).* 69, (2010).

Singh, S. S., Pandey, S. C., Srivastava, S., Gupta, V. S., Patro, B., & Ghosh, A. C. Chemistry and medicinal properties of Tinospora cordifolia (Guduchi). *Indian J. Pharmacol.* 35 (2003), 83-91.

Singh, V., Shukla, S. L. & Mahanwal, J. S. Identification and estimation of morphine in opium samples by derivative U.V. spectroscopy. *J. Indian Forensic Sci.* 28, (1989).

Skundric, D. S., Bealmear, B. & Lisak, R. P. Induced upregulation of IL-1, IL-1RA and IL-1R type I gene expression by Schwann cells. *J. Neuroimmunol.* 74, (1997).

Sneader, W. *Drug Discovery: A History. Drug Discovery: A History* (2006). doi:10.1002/0470015535.

Sobarzo-Sánchez, E. *Alkaloids: Biosynthesis, biological roles and health benefits. Alkaloids: Biosynthesis, Biological Roles and Health Benefits* (2015).

Stavrakov, G., Philipova, I., Lukarski, A., Atanasova, M., Zheleva, D., Zhivkova, Z. D., Ivanov, S., Atanasova, T., Konstantinov, S., & Doytchinova, I. Galantamine-curcumin hybrids as dual-site binding acetylcholinesterase inhibitors. *Molecules* 25, (2020), 3341.

Staerk, D., Lykkeberg, A. K., Christensen, J., Budnik, B. A., Abe, F., & Jaroszewski, J. W. In vitro cytotoxic activity of phenanthroindolizidine alkaloids from Cynanchum vincetoxicum and Tylophora tanakae against drug-sensitive and multidrug-resistant cancer cells. *J. Nat. Prod.* 65, (2002), 1299–1302.

Suzuki, T. & Waller, G. R. Purine alkaloids of the fruits of Camellia sinensis L. and Coffea arabica L. during fruit development. *Ann. Bot.* 56, (1985).

Tafrihi, M., Imran, M., Tufail, T., Gondal, T. A., Caruso, G., Sharma, S., Sharma, R., Atanassova, M., Atanassov, L., Valere Tsouh Fokou, P., & Pezzani, R. The wonderful activities of the genus Mentha: Not only antioxidant properties. *Molecules* 26 (2021), 1118.

Tan, H. L., Chan, K. G., Pusparajah, P., Duangjai, A., Saokaew, S., Mehmood Khan, T., Lee, L. H., & Goh, B. H. Rhizoma coptidis: A potential cardiovascular protective agent. *Front. Pharmacol.* 7 (2016), 362.

Tan, L. T., Lee, L. H., Yin, W. F., Chan, C. K., Abdul Kadir, H., Chan, K. G., & Goh, B. H. Traditional uses, phytochemistry, and bioactivities of Cananga odorata (ylang-ylang). *Evid. based Complement. Alternat. Med.* 2015 (2015), 896314.

Tanaka, J. C. A., da Silva, C. C., de Oliveira, A. J. B., Nakamura, C. V. & Dias Filho, B. P. Antibacterial activity of indole alkaloids from Aspidosperma ramiflorum. *Brazilian J. Med. Biol. Res.* 39, (2006).

Tembo, N., Lampiao, F., Mwakikunga, A. & Chikowe, I. Ethnobotanical survey of medicinal plants used for cervical cancer management in Zomba District, Malawi. *Sci. African* 13, (2021).

Thawabteh, A., Juma, S., Bader, M., Karaman, D., Scrano, L., Bufo, S. A., & Karaman, R. The biological activity of natural alkaloids against herbivores, cancerous cells and pathogens. *Toxins* 11 (2019), 656.

Todd, F. G., Stermitz, F. R., Schultheis, P., Knight, A. P. & Traub-Dargatz, J. Tropane alkaloids and toxicity of Convolvulus arvensis. *Phytochemistry* 39, (1995).

Trask, O. J. *Nuclear Factor Kappa B (NF-κB) Translocation Assay Development and Validation for High Content Screening. Assay Guidance Manual* (2004).

Trompetero, A., Gordillo, A., del Pilar, M. C., Cristina, V. M. & Bustos Cruz, R. H. Alzheimer's Disease and Parkinson's Disease: A Review of Current Treatment Adopting a Nanotechnology Approach. *Curr. Pharm. Des.* 24, (2017).

Tundis, R., Loizzo, M. R., Statti, G. A., Passalacqua, N. G., Peruzzi, L., & Menichini, F. Pyrrolizidine alkaloid profiles of the Senecio cineraria Group (Asteraceae). *Z. Naturforsch. C J. Biosci.* 62, (2007), 467–472.

Turner, M. D., Nedjai, B., Hurst, T. & Pennington, D. J. Cytokines and chemokines: At the crossroads of cell signalling and inflammatory disease. *Biochim. Biophys. Acta Mol. Cell Res.* 1843 (2014).

Umadevi, M., Kumar, K. P. S., Bhowmik, D. & Duraivel, S. Traditionally Used Anticancer Herbs In India. *J. Med. Plants Stud.* 1, (2013).

Vera-Reyes, I., Huerta-Heredia, A. A., Ponce-Noyola, T., Cerda-García-Rojas, C. M., Trejo-Tapia, G., & Ramos-Valdivia, A. C. Monoterpenoid indole alkaloids and phenols are required antioxidants in glutathione depleted Uncaria tomentosa root cultures. *Front. Environ. Sci.* 3, (2015), 27.

Viljoen, A. Transforming Ancient Knowledge into Phytomedicines - a South African Perspective. *Planta Med.* 79, (2013).

Vladimir-Knežević, S., Blažeković, B., Kindl, M., Vladić, J., Lower-Nedza, A. D., & Brantner, A. H. Acetylcholinesterase inhibitory, antioxidant and phytochemical properties of selected medicinal plants of the lamiaceae family. *Molecules* 19, (2014), 767–782.

Wansi, J. D., Devkota, K. P., Tshikalange, E. & Kuete, V. Alkaloids from the Medicinal Plants of Africa. in *Medicinal Plant Research in Africa: Pharmacology and Chemistry.* Elsevier (2013). doi:10.1016/B978-0-12-405927-6.00014-X.

Waterman, P. G. Lignans. Chemical, biological and clinical properties. *Biochem. Syst. Ecol.* 20, (1992).

West, A. P., Abdul, S., Sherratt, M. J. & Inglis, T. J. J. Antibacterial activity of sucralfate against Escherichia coli, Staphylococcus aureus and Pseudomonas aeruginosa in batch and continuous culture. *Eur. J. Clin. Microbiol. Infect. Dis.* 12, (1993).

Wilkinson, D. G. Galantamine: A new treatment for Alzheimer's disease. *Expert Review of Neurotherapeutics* vol. 1 (2001).

Wink, M. A Short History of Alkaloids. in *Alkaloids* (1998). doi:10.1007/978-1-4757-2905-4_2.

Wong, S. K., Lim, Y. Y., Abdullah, N. R. & Nordin, F. J. Assessment of antiproliferative and antiplasmodial activities of five selected Apocynaceae species. *BMC Complement. Altern. Med.* 11, (2011).

Xu, Y., Liu, Y., Li, K., Yuan, D., Yang, S., Zhou, L., Zhao, Y., Miao, S., Lv, C., & Zhao, J. COX-2/PGE2 Pathway Inhibits the Ferroptosis Induced by Cerebral Ischemia Reperfusion. *Mol. Neurobiol.* 59, (2022), 1619–1631.

Xu, Y. Q., Jin, S. J., Liu, N., Li, Y. X., Zheng, J., Ma, L., Du, J., Zhou, R., Zhao, C. J., Niu, Y., Sun, T., & Yu, J. Q. Aloperine attenuated neuropathic pain induced by chronic constriction injury via anti-oxidation activity and suppression of the nuclear factor kappa B pathway. *Biochem. Biophys. Res. Commun.* 451, (2014), 568–573.

Xue, Y., Wang, Y., Feng, D. C., Xiao, B. G. & Xu, L. Y. Tetrandrine suppresses lipopolysaccharide-induced microglial activation by inhibiting NF-κB pathway. *Acta Pharmacol. Sin.* 29, (2008).

Yamasaki, H. Pharmacology of sinomenine, an anti rheumatic alkaloid from Sinomenium acutum. *Acta Medica Okayama* 30 (1976).

Yang, J. H., Li, L., Wang, Y., Zhao, J., Zhang, H., & Luo, S. Two new aporphine alkaloids from Litsea glutinosa. *Helv. Chim. Acta* 88, (2005), 2523-2526.

Yang, L. & Stöckigt, J. Trends for diverse production strategies of plant medicinal alkaloids. *Natural Product Reports* vol. 27 (2010).

Yatoo, M. I., Gopalakrishnan, A., Saxena, A., Parray, O. R., Tufani, N. A., Chakraborty, S., Tiwari, R., Dhama, K., & Iqbal, H. Anti-Inflammatory Drugs and Herbs with Special Emphasis on Herbal Medicines for Countering Inflammatory Diseases and Disorders - A Review. *Recent Pat. Inflamm. Allergy Drug Discov.* 12, (2018), 39-58.

Ye, G., Li, Z., Xia, G., Peng, H., Sun, Z., & Huang, C. A new iridoid alkaloid from the flowers of Plumeria rubra L. cv. acutifolia. *Helv. Chim. Acta* 92, (2009), 2790-2794.

Yuan, X., Tong, B., Dou, Y., Wu, X., Wei, Z., & Dai, Y. Tetrandrine ameliorates collagen-induced arthritis in mice by restoring the balance between Th17 and Treg cells via the aryl hydrocarbon receptor. *Biochem. Pharmacol.* 101, (2016), 87-99.

Zárate, R., Dirks, C., Van der Heijden, R. & Verpoorte, R. Terpenoid indole alkaloid profile changes in Catharanthus pusillus during development. *Plant Sci.* 160, (2001).

Zhan, G., Miao, R., Zhang, F., Chang, G., Zhang, L., Zhang, X., Zhang, H., & Guo, Z. Monoterpene indole alkaloids with acetylcholinesterase inhibitory activity from the leaves of Rauvolfia vomitoria. *Bioorg. Chem.* 102, (2020), 104136.

Zhang, J., Lao, A. & Xu, R. Steroidal alkaloids from Fritillaria thunbergii var. chekiangensis. *Phytochemistry* 33, (1993).

Zhang, Y. T. & Jiang, J. Q. Alkaloids from Aristolochia manshuriensis (Aristolochiaceae). *Helv. Chim. Acta* 89, (2006).

Zhao, L., Wang, L., Di, S. N., Xu, Q., Ren, Q. C., Chen, S. Z., Huang, N., Jia, D., & Shen, X. F. Steroidal alkaloid solanine A from Solanum nigrum Linn. exhibits anti-inflammatory activity in lipopolysaccharide/interferon γ-activated murine macrophages and animal models of inflammation. *Biomed. Pharmacother.* 105, (2018), 606-615.

Zhou, H., Li, J., Sun, F., Wang, F., Li, M., Dong, Y., Fan, H., & Hu, D. A Review on Recent Advances in Aloperine Research: Pharmacological Activities and Underlying Biological Mechanisms. *Front. Pharmacol*. 11 (2020), 538137.

Chapter 4

Alkaloids from the *Solanum* Genus: An Overview

Marcos Venicius Nunes[1], Célio Fernando Figueiredo Angolini[1] and Ana Paula Aparecida Pereira[2,*]

[1]Center for Natural and Human Sciences, Federal University of ABC, Santo Andre, SP, Brasil

[2]Faculty of Nutrition, Federal Unuversity of Mato Grosso, Cuiabá, MT, Brasil

Abstract

The Solanaceae family has a cosmopolitan distribution, widely diversified in number of genera and species, being recognized for its food, medicinal and pharmacological potential. The genus *Solanum* stands out for being the most representative, as well as to produce different alkaloids. In the present review, information related to the biological activities of *Solanum* alkaloids is described. Species such as *Solanum crinitum*, *Solanum lycocarpum*, *Solanum oocarpum*, *Solanum sisymbriifloium*, for example, have alkaloids such as solamargine, solasonine, tomatidine and kukoamines, among others. These compounds have different structures and, consequently, different biological activities, such as: antimicrobial, antilarvicidal, anticancer, antioxidant, antidiabetic, cytotoxic and hepatoprotective potential stand out. Further studies are recommended that can identify novel alkaloids, as well as their biological activities.

Keywords: *Solanum*, alkaloids, biological activity, hepatoprotective, natural products

---

* Corresponding Author's Email: pereira.anap23@gmail.com.

In: The Essential Guide to Alkaloids
Editor: Deepak Kumar Semwal
ISBN: 979-8-88697-456-0

1. Introduction

The plants of Solanaceae family are angiosperms with a cosmopolitan distribution, founded mainly in the Neotropical region. It is widely diversified having 115 different genera and about 3000 described species, some of then being economically recognized for their food, medicinal and pharmacological potential (Gomes, Lima, 2014; Samuels, 2015; Pereira, Rodrigues, Vega, 2016; Dupin et al., 2017; Flora do Brasil, 2020).

In Brazil, 36 genera are described with about 504 species, of which 237 are endemic and are present in different biomes from humid habitat such as the Atlantic Forest and Amazon, as well as arid regions such as the Cerrado and Caatinga (Agra, Silva, Berger, 2009; Samuels, 2015; Flora do Brasil, 2020).

The *Solanum* genus stands out for being the most representative and expressive of the Solanaceae family, especially to its economic importance, but also to its morphological plasticity, diversity, and large number of species with a variety of already reported chemical composition (Olmstead, 2013; Sampaio, Araújo, Agra, 2014; Sinani, Eltayeb, 2017; Sampaio et al., 2019). *Solanum* is described as the largest genus of the family, with about 1700 species spread worldwide (Bohs, 2007). In Brazil, 283 species are described, being 138 endemics, found in the Southeast, South and Northeast regions (Stehmann et al., 2015; Flora do Brasil, 2020).

Regarding the chemical composition, several studies point out to the genus *Solanum* being richiest in diversity of substances having molecules from steroids, saponins, phenolic acids, terpenes, flavonoids, and alkaloids classes (Weissenberg, 2001; Breithaupt, Bamedi, 2002; Sun et al., 2006; Suthar, Mulani, 2008; Ren et al., 2009; Zhang et al., 2010; Milner et al., 2011; Oliveira et al., 2020). Among these classes, *Solanum* stands out for the production of a wide variety of alkaloids, having different biological activities, such as: tropanes, quinolizidines, indolics, pyridines, quinolines, steroids, pyrolizidines, piperidines, isoquinolines (Simões et al., 1999; Pereira, Rodrigues, Veja, 2016).

Depending on the species, the concentration and type of alkaloid can vary, being found in different plant organs, such as leaves, fruits, seeds, roots, or bark. These compounds have different physiological effects, being the basis for the production of several drugs with varied pharmacological and therapeutic activities (Jayakumar, Murugan, 2016; Guimarães, 2018; Ávarez, 2021).

Alkaloids founded in the genus *Solanum* have different known activities, sucha as: antibacterial, against *Staphylococcus aureus*, *Eschirichia coli*, *Pseudomona aeruginosa* and *Salmonella enteritidis* (Palomino, 2014; Ramírez et al., 2017; Cañon, Menco, 2018; Espinoza-Aguikar, Robles-Quispe, 2021); antifungal, against *Candida albicans* (Ferrer-Hernández et al., 2021); larvicidal potential on *Aedes aegypti* (Cruz et al., 2019); preventive and adjuvant agents in the treatment of diabetes (Pereira et al., 2021); anticancer, antitumor and cytotoxic actions (Ahmad et al., 2017; Sinani, Eltayeb, 2017); as well as antioxidant, hypoglycemic and hypolipidemic (Ahmad et al., 2017; Pereira et al., 2019).

In this scenario, it is indisputable that the genus *Solanum* has a great economic role and is a great source of biological compounds. However, we have a great diversity of species from this genus in Brazil not yet studied, so it is imperative that further research to identify and elucidate the function of possible new alkaloids is needed (Lima, Santos, Smozinski, 2014).

Therefore, the mainly objective of this review is to describe the alkaloids compositions founded in species of the genus *Solanum* and their biological activities, presenting a critical-theoretical overview that aims to conduct future scientific investigations to strengthen the field of study.

2. *Solanum* Alkaloids: A General Overview

The *Solanum* genus is recognized for their production of a variety of alkaloids, already identified in more than 100 species, with a wide range of bioactivities of interest for the treatment and prevention of several diseases (Wink, 2003; Munari et al., 2012; Hassine, Mansour, Hammami, 2013; Hernández-Herrera et al., 2014; Diaz, 2015). Alkaloids are characterized as cyclic compounds, which often are derived from amino acids, and containg a nitrogen atom (N) that can exist in the form of a primary, secondary, tertiary, or quaternary amine. However, they may come from terpenes and sterols (Cabral, Pita, 2015; Jayakumar, Murugan, 2016), and have characteristics of weak bases or amphoteric, with general low solubility in water, being mostly colorless and with a bitter taste (Kukula-Koch, Widelski, 2017).

There is no consensus on the exact number of alkaloids elucidated so far. Some authors cite 16,000 (Murphy, 2017), 20,000 (Coqueiro, Verpoorte, 2015) and up to 27,000 thousand (Wink, 2016; Kukula-Koch, Widelski, 2017), but due its broad and general descriptions a lot of molecules can be included on this class and frequently more specific characteristcs need to be used to describe the alkaloids subclasses. For instance, they can be classified according to their biosynthetic pathway as: (I) true alkaloids, derived from amino acids such as L-ornithine, L-tryptophan, and L-histidine, having a nitrogen atom attached to the heterocyclic ring, such as morphine (Figure 1. A); (II) protoalkaloids, which are compounds that have a nitrogen atom derived from amino acids such as L-tyrosine and L-tryptophan, but which are not part of the heterocyclic ring, such as mescaline (Figure 1. B); (III) pseudoalkaloids, which have a skeleton not derived from amino acids, so nitrogen comes from other metabolic pathways (Jayakumar, Murugan, 2016; Dewick, 2009), such as solanidine (Figure 1).

Alkaloids can be produced by different organisms such as bacteria, fungi, marine animals, and plants, and therefore having multiple porpose for those (Bhadra, Kumar, 2011; Amirkia, Heinrich, 2014; Wink, 2016). Plants can synthesize different classes of alkaloids, which have functions such as a reserve of nitrogen, attraction of pollinators, growth regulation and defense against microorganisms and predators, by acting in the process of resistance to pathogens (Dewick, 2009; Cardenas et al., 2015; Pinto et al., 2016; Simões, 2017).

This class of compounds is also recognized for the physiological effects they have on the nervous system of animals and humans, many of which are used as poisons or hallucinogens (Vizzotto, Krolow, Weber, 2010; Jamwal, Bhattacharya, Puri, 2018). Ingestion of some alkaloids, in high concentrations, is associated with seizures, headaches, dizziness, vomiting, among other symptoms. However, these molecules also exert beneficial effects, having several applications such as antioxidants, anti-inflammatory, antibacterial, antitumor, antimalarial, antiparasitic, sedative and analgesic (Funauama, Cordell, 2014; Almoulah et al., 2017). Such harmful and therapeutic effects are due to the chemical and structural diversity of its biologically active components (Moloudizargari et al., 2013; Erharuyia et al., 2014).

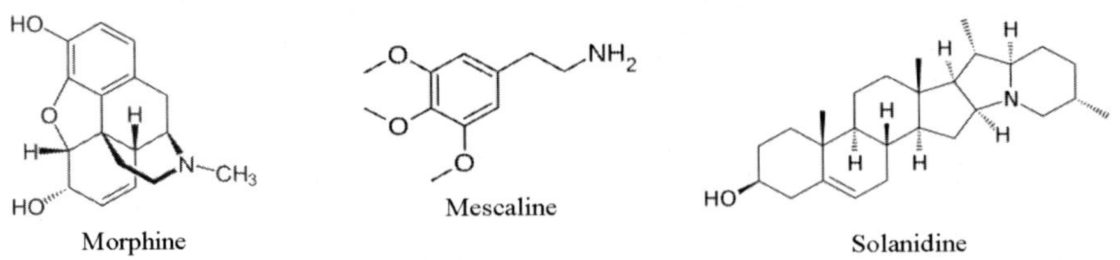

Figure 1. Structures of selected alkaloids isolated from plants.

Table 1. Main alkaloids found in plants of the *Solanum* genus (Solanaceae)

Scientific name	Bioactive compounds	Part	Methods	References
Lobeira-doce (*Solanum crinitum*)	Solamargine 20-epi-Solamargine Solasonine	Green fruit	H and C NMR spectra (including HMQC and HMBC)	Cornelius et al., 2010
Fruta-do-lobo (*Solanum lycocarpum*)	Solasonine Solamargine 12-Hydroxysolasonine Robeneosides A Robeneosides B	Dried fruit	HRFABMS	Yoshikawa et al., 2007
	Solasonine Solamargine	Green fruit	TLC and GC/MS	Araújo et al., 2010
	Unidentified alkaloids	Fruit	Percolation (ethanol) Chromatography plates of silica gel matrix factors (Rfs)	Mendes et al., 2020
	Solasonine Solamargine Hydroxy solamargine isomers Hydroxy solasonine isomers Di-hydroxy solamargine isomers Di-hydroxy solasonine isomers	Pulp, peel, and seeds (unripe and ripe fruit)	UHPLC-Q-ToF	Pereira et al., 2021
Juá-açu (*Solanum oocarpum*)	Kukoamine A Kukoamine B	Pulp	UHPLC-Q-ToF	Pereira et al., 2019
Joá (*Solanum sisymbriifolium*)	Solamargine β-Solamarine	Whole fruit	VLC-TLC	Bagalwa et al., 2010
Chimarrona berinjela (*Solanum torvum*)	Solasodine	Seeds Nodal segment	Murashige and Skoog (MS) basal medium	Moreira et al., 2010
Tomateiro-silvestre (*Solanum caavurana* Vell.)	Caavuranamide 4-Tomatiden-3-one 5 α-Tomatidan-3-one	Fruit	Spectrum IV NMR HRESIMS / MS TLC	Vaz et al., 2012
Jurubeba-de-cipó (*Solanum seaforthianum*)	Solasodine Solanocapsin Solanoforthin 0-Methyl-solanocapsin	Roots Stalks Leaves Fruits	Spectrum IR NMR KBr	Ferrer-Hermández et al., 2016

Table 1. (Continued)

Scientific name	Bioactive compounds	Part	Methods	References
Beladona (*Solanum elaeagnifolium*)	Unidentified alkaloids	Fruits	Qualitative phytochemical tests and fractionated by vacum liquid chromatography (VLC) and monitoring of fractions by thin layer chromatography MTT (cell test) Rp-HPLC-MS	Hernández et al., 2017
Jurubeba-roxa (*Solanum paludosum)*	Solasodine	Whole plant	BOD and UR DMSO	Cruz et al., 2019
Potato (*Solanum phureja*)	α-Solanine α-Chaconine	Peel	ELP ESL HPLC-PDA	Eraso-Grisales et al., 2019
Cusmayllo (*Solanum radicans*)	Unidentified alkaloids	Aerial part	DPPH and FRAP CIM and CBM	Garcia et al., 2020
Jurubeba (*Solanum paniculatum* L.)	Jurubine	Fruit	VLC Spectrum NMR	Vieira et al., 2013
	Unidentified alkaloids	Leaves	Gravimetric method Drying Spectrophotomet	Marttins et al., 2021
Tomate-do-mato (*Solanum pimpinellifolium* L.*)*	Unidentified alkaloids	Leaves	CCD (f_{254}) UV chamber BDO	Fioresi et al., 2021

Although many studies have already been carried out to verify and validate the properties and applications of *Solanum* species and their metabolites, many of them still have their chemical composition of alkaloids unknown or superficially studied (Kaunda, Zhang, 2019; Souto, Silva, 2020). The present review focuses on discussing the biological properties and applications of the alkaloids already characterized, described in Table 1.

2.1. Steroidals Alkaloids and Steroidals Glycoalkaloids

Steroids and glycoalkaloids are naturally synthesized by the genus *Solanum*, being reported in about 350 species such as potato, tomato and eggplant, and by other members of the Solanaceae family. They have a chemical defense function against the attack of fungi, nematodes, herbivores, as well as they can act in response to adverse conditions of environmental stress (Friedman et al., 2009; Friedman, 2015; Chowanski et al., 2016; Castro, Pereira, Santos, 2017). In addition, they are of therapeutic interest due to their cytotoxic, anticancer, anti-inflammatory, antifungal and molluscicidal activities (Ikeda, Tsumagari, Nohara, 2003; Jarald et al., 2008; Silva et al., 2008; Pinto et al., 2011).

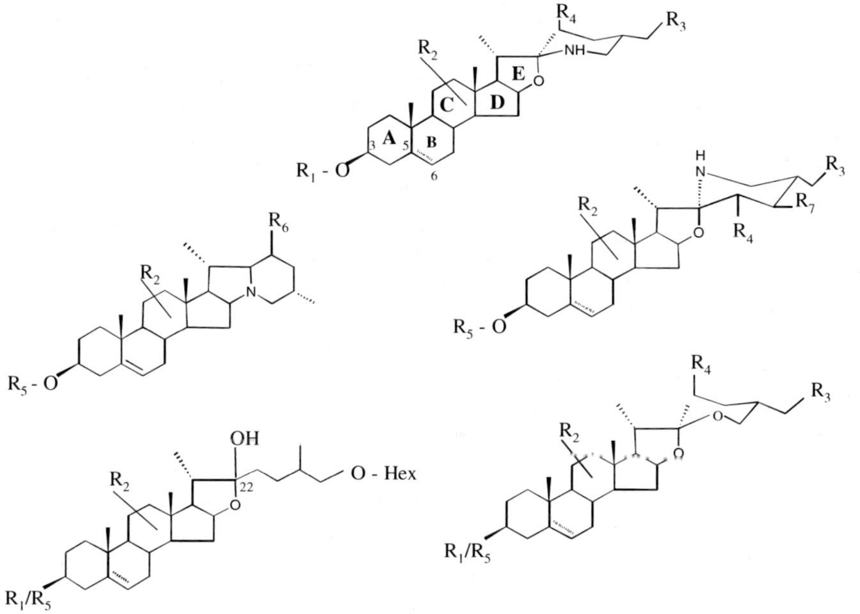

Figure 2. General structure of Solanaceae steroid glycosides. The molecules of these two classes consist of a pentacyclic steroidal skeleton, a modified side chain (cyclic, open, nitrogen-containing in the case of alkaloids) originated from the cholesterol side chain and are glycolyzed in the OH group at C_3 (R1, 5). Other hydroxylations, acetylations and O-glycosylations can occur in the backbone or side chain. R1: 2*Hex – 2*Pent, Pent – deoxHex, 3*Hex – (Pent)0-2, Pent – deoxHex, deoxHex, 4*Hex – (Pent)0-1, 5*Hex – (Pent)0-1; R2: H, OH, OMe, OAc; R3: H, OH, O-Hex, O-Pent, O-Hex-Pent; R4: H, OH, OAc; R5: Hex – (deoxHex)1-2, 2*Hex – (deoxHex)1-2, 4*Hex; R6: H, OH, OAc; R7: H, OH, O-Hex. Hex: hexose, Pent: pentose, deoxHex: desoxihexose.

Their concentrations are generally higher in parts of the plant with a high rate of metabolism, such as young leaves and flowers, showing a consequent decrease with low metabolic activity, as well as with the ripening of the fruit (Friedman, 2006; Pereira et al., 2021). In general, Solanaceae steroid glycosides consist of three main structural parts: (I) a pentacyclic steroidal backbone (marked A-E rings) that may contain a double bond at C_5 (Figure 2); (II) a modified side chain derived from the C_{20}-C_{27} cholesterol side chain (Figure 2); and (III) a glycosyl moiety at the C_3 hydroxyl group (Figure 2). These classes of alkaloids have a wide variety of compounds, being tomatidine, solasodine and solanine, furostanol and spirostanol the main aglycone structures commonly found in the Solanaceae family (Heinig & Aharoni, 2014).

These compounds have already been isolated from different species of the *Solanum* genus (Cornelius et al., 2010; Chauhan et al., 2011) and have their biological activities investigated and applied in therapeutic porposes. Because it is a group characteristic of the *Solanum* genus and cover a variety of chemical compounds such as solasonine, solamargine, solasodine, tomatidines, jurubines and some of their derivatives; the next topics will discuss the main glycoalkaloids, their aglycone forms and derivatives, as well as their biological activities.

2.1.1. Solasodine Type Alkaloids

2.1.1.1. Solasonine and Solamargine

Solasonine and solamargine have its aglycone solasodine in common, however, they are distinguished by the sugar chain associated with the position of carbon 3, which are a triose that in solasonine is solatriose (Figure 3.A), composed of rhamnose and glucose linked at positions 2 and 3 of galactose, while in solamargine is as chacotriose (Figure 3. B), consists of rhamnose linked to carbons 2 and 4 of glucose (Weisserberg, 2001; Simões et al., 2002; Soares-Mota et al., 2010; Eraso-Grisales, Mejía-España, Hurtado-Benavides, 2019).

They are present in several species of the genus *Solanum* (Table 1) and have different biological activities. In human health, for instance, crude extracts obtained from *Solanum lycocarpum* fruits showed antiproliferative activities against murine melanoma (B16F10), human colon carcinoma (HT-29), human breast adenocarcinoma (MCF-7) tumor cell lines, human cervical adenocarcinoma (hepatocellular carcinoma of the human liver - HepG2) and human glioblastoma (MO59J, U343, and U251), showing better results for HepG2 (Munari et al., 2014).

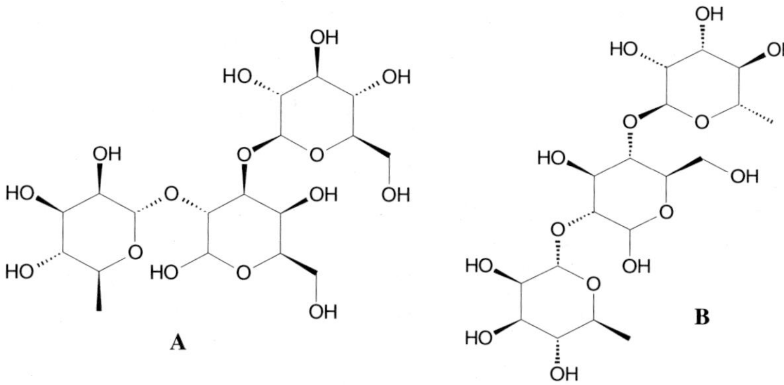

Figure 3. Structure of solanidine trioses. (A) solatriosis and (B) chacotriosis.

Other study performed with free alkaloid extract and encapsulated by Nanostructured Lipid Carriers (NCL) from *S. Lycocarpum* verified the cytotoxicity of solasonine and solamargine in RT4 bladder cancer cell line. It was observed that, after 24 hours, the free extract showed a greater cytotoxic effect, while the extract encapsulated in CLN was more cytotoxic after 48 and 72 hours. This action time behavior is associated with the solasonine and solamargine release profile of the nanoparticles. The study highlights the compounds as future bladder cancer drugs (Carvalho, 2016; Carvalho et al., 2019).

Alkaloids present in ethanolic extracts of leaves and fruits of *Solanum asperum* show antifungal activity against clinical isolates of *Trichophyton rubrum*, *T. mentagrophytes*, *Microsporum canis* and *M. gypseum* and in yeasts of *Candida albicans* ATCC 10231 and *C. parapsilosis* ATCC 22019 for determination of the bioactivity level of the extracts, the Minimum Inhibitory Concentration (MIC) was verified. It was observed that solamargine presented values of 125 μg/mL for yeast and 62.5 μg/mL for filamentous fungus, while solasonine was more active, with values of 62.5 μg/mL for yeast and 0.24 μg/mL for filamentous fungus. In view of these results, these authors point out thaht these compounds may be an alternative for the treatment of mycoses in humans (Pinto et al., 2011).

About their antiparasitic activity, these compounds have a positive action in the fight against flagellated protozoa parasites of humans, highlighting their effectiveness in the control of *Giardia lamblia*, the cause of giardiasis. In the study, a crude extract (EB 96% ethanol) of *S. Lycocarpum* fruits was prepared and then partition fractionated (EF 40% ethanol and n-hexane: ethyl acetate). The alkaloids were then isolated by column chromatography, and five extracts were made: EB, EF, solasonine, solamargine and a mixture of these compounds (1:1). Inhibitory Concentration (IC50) was observed, and all extracts showed activity against *G. Lamblia* ([IC50]: EB = 105.4; EF = 95.0; solasonine = 103.7, solamargine = 120.3 and the mixture of alkaloids 1:1 = 13.23 μg/mL). It is worth noting that the mixture (1:1) was more active when compared with each alkaloid in its individual form, showing a synergy of the two compounds (Martins et al., 2015).

In another study, the antioxidant capacity of extracts obtained from seeds, pulp and peel of unripe and ripe fruits of *S. lycocarpum* and applied in Oxygen Radical Absorption Capacity assays [lipophilic ORAC (ORAC-L) and hydrophilic ORAC (ORAC-H)] was observed, as well as changes in the compounds during fruit ripening by means of High-Performance Liquid Chromatography (UHPLC) coupled to a Mass Spectrometer (MS). Statistical analysis showed that ORAC-H values were more expressive in unripe fruit seeds (1256.75 ± 4.70 μmol Trolox equivalent/mg) and ripe fruit seeds (1145.89 ± 0.89 μmol Trolox equivalent/mg). mg), highlighting the antioxidant action of the compounds presents in those parts and ripening stages (Pereira et al, 2021), however in those extract alkaloid and phenolic compounds were present, and the author were not able do distinguish the source of antioxidant activity.

2.1.1.2. Solasodine

Solasodine (Figure 4) is the aglycone of solasonine and solamargine. It has high solubility in organic solvents such as benzene and chloroform and low solubility in methanol, acetone, and ethanol. Its lipophilic characteristic can be evidenced by the presence of the steroid nucleus in its structure (Santos, 2013). As it is an aglycone, i.e., without sugar, solasodine has a reduced surface-active property, reducing its ability to lyse cell membranes and, consequently, being less cytotoxic than its glycosylated versions.

In vitro and *in vivo* tests showed its neurogenic action in rats and mice, in which solasodine demonstrated activity for inducing teratocarcinoma (P19), precursor cells (NT2) and pheochromocytoma (PC12) cells to differentiate into new neurons. Differentiated P19 cells were observed to exhibit a cholinergic phenotype. This solasodine-induced neuronal differentiation process was confirmed by the expression of DCX, a marker that has been shown to be specifically expressed in newly formed neuroblasts. These results point to solasodine as a drug to induce neurogenesis as part of neuronal replacement therapy (Lecanu et al., 2011).

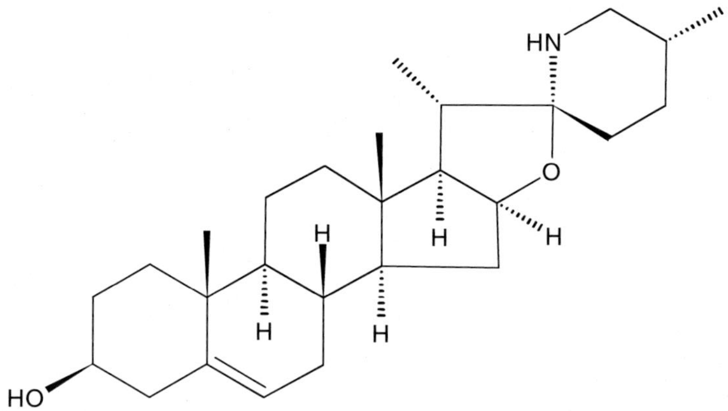

Figure 4. Structure of solanidine.

In an experimental study, solasodine was isolated from *Solanum sisymbriifolium* fruits and tested for its anticonvulsant and central nervous system (CNS) depressant effect in mice. Seizures were induced by pentylenetetrazol (PTZ) and picrotoxin (PCT) and sleep time induced by thiopental. The results showed that solasodine significantly ($p<0.01$) delayed the latency of the tonic extensor phase of the hind limbs in PTC induced seizures but was not efficient for PTZ. The treatment significantly potentiated ($p<0.001$) the sleep caused by thiopental. Thus, the study reports the anticonvulsant and CNS depressant activity of solasodine (Chauhan et al., 2011).

Another study was conducted to investigate its effect on scopolamine-induced amnesia and brain cholinesterase activity in young and old mice. Doses of solasodine (1, 2 and 4 mg/kg, p.o) were administered and it was observed that there was a reduction ($p<0.05$) in the transfer latency (TL) in the young and the elderly. Cholinesterase activity was also reduced in aged mice. The probable mechanism of solasodine can be attributed to the facilitation of cholinergic transmission and, therefore, to the improvement of memory. These results indicate the therapeutic potential of solasodine in the management of patients with cognitive disorders (Alpesh et al., 2011; Desai et al., 2011).

Another use of solasodine is associated with its antimicrobial potential. Methanolic extracts from *Solanum leucocarpum* samples were tested for antibacterial activity against *Staphylococcus aureus* (ATCC No. 6538) and an MIC action of 62.5 μg/mL was observed (Niño, Correa, Mosquera, 2009).

2.1.2. Tomatidine Type Alkaloids

Tomatidine (Figure 5) is the most abundant glycoalkaloid in *Solanum lycopersicum*. This alkaloid is the aglycone of α-tomatine, which has the original molecule bound to a tetrasaccharide composed of a xylose, a galactose and two glucoses (Andrino, 2010).

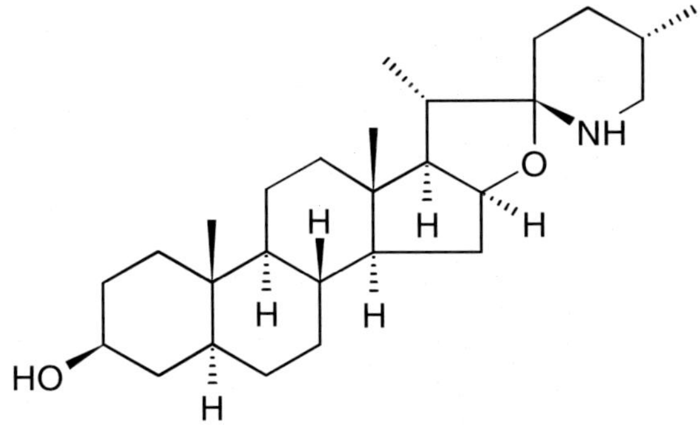

Figure 5. Tomatidine structure.

Tomatidine obtained from aqueous and alcoholic extracts from powdered dry leaves of *Solanum lycopersicum* showed low molluscicidal activity against the *Biomphalaria glabrata,* evaluating the lethal concentrations (CL), it was observed that tomatidine presented LC50 and LC90 of 4.49 ppm and 11.84 ppm, respectively (Leyton et al., 2005).

In a study to evaluate the antiviral activity of tomatidine, *in vitro* tests were performed with this alkaloid at 10 μM in human liver cell lines Huh7 (JCRB0403) and Vero-WHO African green monkey kidney cells, which were infected with Chikungunya (CHIKV) in the multiplicity of infection (MOI). It was observed that EC50 values were 1.3 μM for CHIKV. In this study, the authors sought to identify which stage of the virus replication cycle can tomatidine controls infectivity. Tomatidine was found to act in stages of infection after the entry of CHIKV RNA into Huh7 cells. The expression of genomic (gRNA) and subgenomic (sgRNA) RNA was reduced with tomatidine treatment, as the intensity of the gRNA band was reduced by 1.7-fold (40%) and sgRNA 2-fold (50%). Additionally, it was verified, by Western blot, that tomatidine affected the expression of viral proteins nsP2, E1 and capsid (C) in 0.5 ± 0.1 (50%), 0.18 ± 0.1 (18%) and 0.46 ± 0.1 (46%), respectively, after 9 hours post-infection (hpi). In the 4 hpi pattern, a reduction in protein expression was observed by 0.5 ± 0.1 (50%) for nsP2, 0.3 ± 0.1 (30%) for E1 and 0.5 ± 0.2 (50%) for C. And at 6 hpi the expression was 0.6

± 0.2 (60%) for nsP2, 0.4 ± 0.2 for E1 and 0.6 ± 0.1 for C. In conclusion, tomatidine interferes with the infectivity and expression levels of CHIKV viral proteins, showing pharmacological potential and antiviral efficacy (Troost-Kind et al., 2021).

α-Tomatine is found in leaves, roots, stems, flowers and in higher concentration in unripe tomato fruits. However, its concentration may vary according to the plant development stage, growth conditions and growing region (Kozukue, Friedman, 2003; Friedman, 2004).

The antifungal activity observed when the fungus *Fusarium oxysporum* was exposed to α-tomatine suggests that this glycoalkaloid activates the signaling pathways of phosphotyrosine kinase and monomeric G protein, leading to an increase in Ca^{2+} and accumulation of reactive oxygen species (ROS), which leads the fungus to cell death (Ito et al., 2007).

The anti-trypanocidal activity of α-tomatin was reported in an experiment that observed that concentrations of 10^{-4} and 10^{-5} M of this compound were active over the 50% lethal dose (LD50) of the parasites. To verify the cultures after exposure to the alkaloid for 30 minutes, the samples were fixed with methanol and stained with Giemsa and it was not possible to verify the presence of the parasites, indicating that α-tomatine caused the lysis of its cell wall, evidencing its anti-trypanocidal activity (Evangelista et al., 2018). Also, in a study carried out using crude extracts of leaves and unripe and ripe fruits of *Solanum lycopersicum*, it was possible to observe the antifungal activity of tomatidine and α-tomatin against *Moniliophthora perniciosa* (Andrino, 2010).

2.2. Spermine and Spermidine Cinnamic Acid Derivatives (Kukoamines and Allies)

Cinnamic acid comprises a series of trans-phenyl-3-propenoic acids that differ in their ring substitution. In its chemical structure there are nine carbon atoms, composed of a benzene ring (Liu et al., 1995; Kasetti et al., 2012). Spermine and spermidine are polyamines involved in cellular metabolism that is found in all eukaryotic cells, having various metabolic functions within organisms. Both molecules compose another alkaloid class often found in *Solanum* genus (kukoamines).

2.2.1. Kukoamines

The kukoamines are described as a class of polyamines that have a polymethylenepolyamine skeleton (putrescine, spermidine or spermine) and at least one dihydrocaffeic acid fragment. In general, they are found in several Solanaceae plants such as potato, tomato and tobacco showing antioxidant activity due their cinnamic acid part (Wu et al., 2016; Huang et al., 2017).

Kukoamines A and B are isomers of each other, differing only in the position of dihydrocaffeoylation. While kukoamine A (Figure 6.A) has a linear chain structure in which two dihydrocapheoyl moieties are individually connected to two terminal N-atoms, kukoamine B (Figure 6.B) has one of the dihydrocafeoyl moieties attached to a medium nitrogen atom, forming a branched-chain structure (Li et al., 2018). These two compounds are the most common characterized, however a variety of other strucutures in this class can ben detected however fewer studies were found in literature, point out that these compounds need to be more explored in *Solanum* species.

The structural diversity of kukoamines is due to the ability of polymethylenepolyamine to bind to phenolic molecules. For this reason, the existence of several analogues is also possible, since the amino groups can conjugate at different positions on the polyamide backbone (Li et al, 2015). The *Solanum* genus is known to have kukoamines, but different from steroidal alkaloids, kukoamines are less frenquently, however several species have demonstrated ability to produce them (Table 1). Examples include potato (*Solanum tuberosum*), which contains kukoamine A and several isomers of spermines and spermidines (Parr et al, 2005) and tomato (*S. lycopersicum*) which contains kukoamine C (D'Esposito et al, 2017).

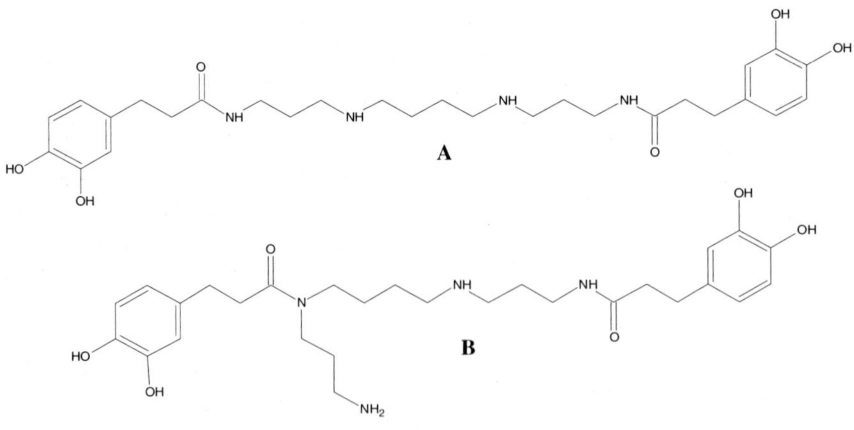

Figure 6. Structure of Kukoamines. (A) Kukoamine A, (B) Kukoamine B.

2.2.1.1. Kukuaminas A and B

Kukoamines A and B were found in *Solanum oocarpum* fruits by Pereira et al., (2021). It was also observed that the extract enriched with these compounds was able to protect mitochondrial function and glucose-induced insulin secretion against damage caused by high exposure to glucose. These results raise the possibility of using such compounds for the prevention or adjuvant therapy of diabetes (Pereira et al., 2021).

Li et al., (2018) investigated the antioxidant and cytoprotective activity in bone marrow-derived mesenchymal stem cells (bmMSCs) under oxidative stress induced by H_2O_2 and Fe^{2+} in rats. They observed that there was a restoration of cell viability when incubated with kukoamines A and B at 56.5-118.4 μM, suggesting that these alkaloids have cytoprotective effects on bmMSCs, making them therapeutic candidates and for clinical applications in nervous system diseases.

In another study conducted with kukoamine A, it was possible to observe that this compound presents itself as a recombinant and physiological furin inhibitor in U-251 MG human glioblastoma cell lines. IC_{50} was calculated with a 95% confidence interval. The IC_{50} for recombinant furin was observed to be 1.07 μM (95% CI = 0.0817-4.577), while the IC_{50} for physiological furin was 193.2 μM (95% CI = 165.5-229). In Molecular Dockin tests, kukoamine A blocked access to the furin active site through hydrogen bonds and hydrophobic interactions, thus inhibiting the catalytic action of furin (Zaragoza-Huesca et al., 2022).

Researchers also investigated the neuroprotective action of kukoamine A on SH-SY5Y human neuroblastoma cells. A model of cell injury induced by H_2O_2 was used to study the ability of anti-oxidative stress, as well as a model of Ca^{2+} overload induced by N-methyl D-aspartate (NMDA) to verify the blocking capacity of this compound. The results showed that kukoamine A attenuated cellular apoptosis induced by H_2O_2 and intracellular Ca^{2+} overload, in addition to modulating the expression of Bax, Bcl-2 and p53 proteins, procaspase-3/9 related to apoptosis in SH-SY5Y human neuroblastoma cells, indicating that this alkaloid has anti-oxidative stress capability (Hu et al., 2015).

An experiment in mice was conducted to assess whether kukoamine B would be able to act as a dual inhibitor of lipopolysaccharides (LPS) and oligodeoxynucleotides containing CpG motifs, pathogenic molecules responsible for inducing sepsis. The alkaloid was isolated and identified through the coupling of affinity biosensor with chromatography and the interaction with the targets was evaluated by neutralization assay in cellular assay with macrophages. Kukoamine B showed affinity and neutralized LPS and CpG DNA, preventing their interaction with macrophages. Furthermore, it inhibited the expression of pro-inflammatory mediators (Liu et al., 2010).

Obesity is associated with chronic diseases such as diabetes, cardiovascular problems, hypertension and neurodegenerative diseases. Recently, researchs conducted with kukoamine B investigated its antidiabetic (high fat/fructose diet-induced insulin resistance – HFDRFr), antioxidant and anti-inflammatory properties in rats. The animals were fed with HFDFr for 10 weeks and treated with kukoamine B from the sixth week onwards. Kukoamine B has been observed to attenuate body weight, insulin resistance, lipid accumulation, oxidative stress and inflammation, and this alkaloid is indicated against induced metabolic disorders (Zhao et al., 2020).

Conclusion

Alkaloids are widely present in plants of the Solanaceae family, which have different structures and, consequently, different biological activities, highlighting the genus importance for different therapeutical and medicinal use.

The data presented summarize the chemical compositions of alkaloids from the *Solanum* genus, as well as information on the biological activities to alleviate problems such as diabetes, different types of cancer and neurological problems. Other less frequently applications also suggest its importance in antioxidant, anti-parasitic, anti-microbial and cytotoxic action.

Despite all these activities reported, especially ins diabetes and neuroprotectivy effecs, most of the studies are *in vitro* or in begning of animal assays, showing that further studies are needed to confirm and discovery new functions of those molecules. Also, fewer studies try to elucidate the mechanism of action of these molecules in their different approaches, especially in clinical trials. Adittionaly, as demonstrate most of alkaloids have several positional isomers and can be also present in aglicone or glicosilated form, manking space for several studies to correlate structure and activity (structure activity relationship). Therefore, there is a large area of research to be explored and directly applied in human health and related areas.

References

Agra, M. F., Silva, K. N., Berger, L. R. (2009). Flora da Paraíba, Brasil: *Solanum* L. (Solanaceae) [Flora of Paraíba, Brazil: Solanum L. (Solanaceae)]. *Acta Botânica Brasílica*, 23(3), 826–842. 2009.

Ahmad, R., Fatima, A., Srivastava, A. N., Khan, M. A. (2017). Evaluation of apoptotic activity of *Withania coagulans* methanolic extract against human breast cancer and Vero cell lines. *Jornal of Ayurveda and Integrative Medicine*, 8(3), 177-183.

Almoulah, N. F., Voynikov, Y., Gevrenoca, R., Schohn, H., Tzanova, T., Yagi, S., Thomas, J., Mignard, B., Ahmed, A. A. A., El Siddig, M. A., Spina, R., Laurain-Mattar, D. (2017). Antibacterial, antiproliferative and antioxidante activity of leaf extracts os selected Solanaceae species. *South African Journal of Botany*, 112, 368-374.

Alpesh, B. D., Virendra, G. K., Hanumanthachar, J., Amit, T. R., Harshad, M. (2011). Evaluation of Antiamnesic effect of Solasodine in Mice. *International Journal of PharmTech Research*, 3(2), 732-740.

Álvarez, C. M. C. (2021). Evalución de la presencia de alcaloides em tres estados fenológicos del fruto de tomate de arbol (*Solanum betaceaum*) injerto em palo bobo (*Nicotiana glauca*), em tungurahua. 62f. Proyeucto de Investigación. (Grado de Ingeniera Agrónoma) – Universidad Técnica de Ambato, Facultad de Ciencias Agropecuarias [Evaluation of the presence of alkaloids in three phenological stages of tree tomato fruit (Solanum betaceaum) grafted on palo bobo (Nicotiana glauca), in tungurahua. 62f. Research Project. (Degree of Agricultural Engineering) – Technical University of Ambato, Faculty of Agricultural Sciences].

Amirkia, V., Heinrich, M. (2014). Alkaloids as drug leads-a predictive structural and biodiversity-based analysis. *Phytochemistry Letters,* 10, 48-53.

Andrino, F. G. (2010). Estudo químico de compostos de *Solanum lycopersicum* com atividade antifúngica para *Moniliophthora perniciosa*. 70 f. Tese (Mestrado em Ciências), Escola Superior de Agricultura "Luiz de Queiroz," Universidade de São Paulo [*Chemical study of Solanum lycopersicum compounds with antifungal activity against Moniliophthora perniciosa*. 70 f. Thesis (Master of Science), Higher School of Agriculture "Luiz de Queiroz," University of São Paulo].

Araújo, M. G. F., Galeane, M. C., Castro, A. D., Salgado, H. R. N., Almeida, A. E., Cunha, W. R., Veneziani, R. C. S., Moreira, R. R. D. (2010). Pharmacognostical evaluation of fruits of *Solanum lycocarpum* A. St.-Hill. (Solanaceae). *Pharmacognosy Journal,* 2(9), 248–253.

Bagalwa, J.-J. M., Voutquenne-Nazabadioko, L., Sayagh, C., & Bashwira, A. S. (2010). Evaluation of the biological activity of the molluscicidal fraction of *Solanum sisymbriifolium* against non target organisms. *Fitoterapia*, 81(7), 767–771.

Bhadra, K., Kumar, G. S. (2011). Therapeutic potential of nucleic acid-binding isoquinoline alkaloids: binding aspects and implications for drug design. *Medicinal Research Reviews,* 31(6), 821-862.

Bohs, L. (2007). Major clades in *Solanum* based on ndhF sequence data. In: A Festschrift for William G.'Arcy. *Missouri Botanical Garden*, Saint Louis, 27-49.

Breithaupt, D. E., Bamedi, A. J. (2002). Carotenoids and carotenoid esters in potatoes (*Solanum tuberosum* L.): new insights into an ancient vegetable. *Agricutural and Food Chemical,* 50(24), 7175-7181.

Cabral, C., Pita, J. R. (2015). Alcalóides – Relevância na Farmácia e no Medicamento. *Património Histórico-Farmacêutico* [*Relevance in Pharmacy and Medicine. Historical-Pharmaceutical Heritage*]. Coimbra, 32p.

Cañon T., Menco M. Estudio fitoquímico de la especie vegetal *Solanum crinitipes* Dunal (Solanaceae) y evaluación de uso como agente antimicrobiano. (2018). 76f. [Tesis para optar al Título Profesional de Químico Farmacéutico]. Bogotá: Universidad de Ciencias Aplicadas y Ambientales [Phytochemical study of the plant species Solanum crinitipes Dunal (Solanaceae) and evaluation of its use as an antimicrobial agent. (2018). 76f. [*Thesis to opt for the Professional Title of Pharmaceutical Chemist*]. Bogotá: University of Applied and Environmental Sciences].

Carvalho, I., P., S. (2016). Desenvolvimento de nanopartículas lipídicas sólidas no carreamento de extrato alcaloídico de *Solanum lycocarpum* e avaliação biológica *in vitro* em células de câncer de bexiga. 83f. Dissertação. (Mestrado em Ciências Farmacêuticas) – Universidade de São Paulo [*Development of solid lipid nanoparticles in the carrier of Solanum lycocarpum alkaloid extract and in vitro biological evaluation in bladder cancer cells. 83f. Dissertation.* (Master in Pharmaceutical Sciences) – University of São Paulo].

Castro, C. M., Pereira, A. da S., Santos, A. C. A. (2017). Teor de Glicoalcaloides de Tubérculos de Batata (*Solanum tuberosum* L.). *Comunicado Técnico*, 351. Embrapa Clima Temperado. 1ª edição [*Glycoalkaloids Content of Potato Tubers* (Solanum tuberosum L.). Technical Communiqué, 351. Embrapa Temperado Climate. 1st edition]. 1-4p.

Chauhan, K., Sheth, N., Ranpariya, V., Parmar, S. (2011). Anticonvulsant activity of solasodine isolated from *solanum sissymbriifolium* fruits in rodents. *Pharmaceutical Biology*, 49(2), 194-199.

Chowanski, S., Adamski, Z., Marciniak, P., Rosinski, G., Buyukguzel, E., (2016) A review of bioinsecticidal activity of Solanaceae alkaloids. *Toxins (Basel)* 8(3), 60-70.

Cornelius, M. T. F., Alves, C. C. F., Silva, T. M. S., Alvez, K. Z., Carvalho, M. G., Braz-Filho, R., Agra, M. F. (2004). Solasonina e flavonóides isolados de *Solanum crinitum* Lam [Solasonine from flavonoids isolated from Solanum crinitum Lam.]. *Revista Brasileira de Farmacologia* [*Brazilian Journal of Pharmacology*], 85(2), 57-59.

Cornelius, M. T. F., Carvalho, M. G. de, Silva, T. M. S. da, Alves, C. C. F., Siston, A. P. N., Alves, K. Z., Sant'Anna, C. M. R., Neto, M. B., Eberlin, M. N., Braz-Filho, R. (2010). Other chemical constituents isolated from *Solanum crinitum* Lam. (Solanaceae). *Journal of the Brazilian Chemical Society*, 21(12), 2211–2219.

Cruz, I. L. S., Pardal, B. M., Alves, S. P., De Medeiros, G. A. M., Barbosa-Filho, J. M., Maleck, M. (2019). Alcaloide esteroidal, substância de *Solanum paludosum,* com atividade larvicida sobre *Aedes aegypti* [Steroidal alkaloid, substance from Solanum paludosum, with larvicidal activity against Aedes aegypti]. *Revista de Saúde* [*Health Magazine*], 10(1), 15-19.

D'esposito, D., Ferrielo, F., Molin, A., Diretto, J., Sacco, A., Minio, A., Barone, A., Monaco, R., Cavella, S., Tardella, L., Giuliano, G., Delledone, M., Frusciante, L.,

Ercolano, M. R. (2017). Unraveling the complexity of transcriptomic, metabolomic and quality environmental response of tomato fruit. *BMC Plant Biology*, 17(66).

Dandurand, L. M., Zasada, I. A., LaMondia, J. A. (2019). Effect of the trap crop, *Solanum sisymbriifolium,* on *Globodera pallida, Globodera tabacum,* and *Globodera ellingtonae. Journal of Nematology*, 51(1), 1-11.

De Mendonca, J. L., Lopes, J. F. (2019). Coleção de germoplasma de espécies silvestres de *Solanum* da Embrapa. Brasília: *Embrapa Hortaliças* [*Embrapa's collection of germplasm from wild species of Solanum.* Brasilia: Embrapa Vegetables]. 168p.

Desai, A. B., Kagathara, V. G., Joshi, H., Rangani, A. T., Mungra, H. (2011). Evaluation of Antiamnesic effect of solasodine in mice. *International Journal of PharmTech Research* 3(2), 732-740.

Dewick, P. M. (2009). *Medicinal Natural Products: a biosynthetic approach*. London: West Sussex John Wiley & Sons, 3º. Ed., 539p.

Diaz, G. J. (2015). Toxicosis by plant alkaloids in humans and animals in Colombia. *Toxins,* 7(12), 5408-5416.

Dupin, J., Matzke, N. J., Sarkinen, T., Knapp, S., Olmstead, R. G., Bohs, L., Smith, S. D. (2017). Bayesian estimation of the global biogeographical history of the Solanaceae. *Journal of Biogeography,* 44(4), 887-899. 2017.

Eraso-Grisales, S., Mejía-España, D., Hurtado-Benavides, A. (2019). Extracción de glicoalcaloides de papa nativa (*Solanum phureja*) variedad ratona morada con líquidos presurizados [Extraction of glycoalkaloids from native potato (Solanum phureja) purple mouse variety with pressurized liquids]. *Revista Colombiana de Ciencias Químico-Farmacéuticas* [*Colombian Journal of Chemical-Pharmaceutical Sciences*] 48(1), 181-197.

Erharuyia, O., Faloduna, A., Langer, P. (2014). Medicinal uses, phytochemistry and pharmacology of Picralima nitida (Apocynaceae) in tropical diseases: A review. *Asian Pacific Journal of Tropical Medicine,* 7(1), 1-8.

Espinoza-Aguikar, G. R., Robles-Quispe, P. I. Efecto antibacteriano *in vitro* del extracto etanolico de la cáscara de *Solanum sessiliflorum* Dunal (cocona) frenet a las cepas de *Salmonella enteritidis* y *Staohylococcus aureus.* (2021). 58f. Tesis (Tesis para optar el título profesional de Químico Farmacêutico) – Universidad María Auxiliadora, Facultad de Ciencias de la Salud, Escuela Profesional de Farmacia y Bioquímica [In vitro antibacterial effect of the ethanolic extract of the shell of Solanum sessiliflorum Dunal (cocona) against Salmonella enteritidis and Staohylococcus aureus strains. (2021). 58f. Thesis (*Thesis to opt for the professional title of Pharmaceutical Chemist*) – María Auxiliadora University, Faculty of Health Sciences, Professional School of Pharmacy and Biochemistry].

Ferrer-Hernández, A. E., Pérez-Martínez, C., Ortiz-del-Toro, P., Fuentes-Fiallo, V., Teixeira-de Souza Jr, P., Dall-Oglio, E. L., Adriano-Parizotto, C., Basterrechea-Rey, M., Hernández-Escalona, M., Torres-ferrer, M., Aiardes-Ferrer, M. E. (2016). O-23 Metil solanocapsina, un nuevo alcaloide steroidal del *Solanum seaforthianum* Andr [O-23 Methyl solanocapsin, a new steroidal alkaloid from Solanum seaforthianum Andr]. *Revista Cubana de Química* [*Cuban Journal of Chemistry*], 28(3), 890-901.

Ferrer-Hernández, A. E., Teles, C. B., Gonçalves, A. P. P., Cunha, P. A., Lima, N. C. Da S., Ferrer, M. E. A. (2021). Preliminary Phytochemical Study of *Solanum crinitum*

Lam. (Solanaceae family) and Analysis of its Microbiological Activity. *Revista Cubana de Plantas Medicinais* [*Cuban Journal of Medicinal Plants*], 26(1), 1-16.

Fioresi, R. S., Rodrigues-Filho, J., Perin, I. T. A. L. da Silva, R. W., dos Santos, C. R., Corte, V. B., França, H. S. (2021). Allelopathic effect of *Solanum pimpinellifolium* L. on the germination and initial growth of *Lactuca sativa* and *Bidens pilosa. Scientia Plena* [*Full knowledge*], 17(6), 1-10.

Flora do Brasil. (2020). [Internet]. *Jardim Botânico do Rio de Janeiro* [*Botanical Garden of Rio de Janeiro*]. Available in: http://floradobrasil.jbrj.gov.br/reflora/floradobrasil/FB225.

Friedman, M. (2006). Potato glycoalkaloids and metabolites: Roles in the plant and in the diet. *Journal of Agricultural and Food Chemistry*, 54(23), 8655-8681.

Friedman, M. (2015). Chemistry and anticarcinogenic mechanisms of glycoalkaloids produced by eggplants, potatoes, and tomatoes. *Journal of Agricultural and Food Chemistry*, 63, 3323-3337.

Friedman, M., Levin, C. (2009). Analysis and biological activities of potato glycoalkaloids, calys tegine alkaloids, phenolic compounds, and anthocyanins, En: "*Advances in Potato Chemistry and Technology*," Ed por J. Singh, y L. Kaur, Elsevier, Burlington MA, Massachusetts, 127-161p.

Funayama, S., Cordell, G. A. (2014). *Alkaloids: A Treasury of Poisons and Medicines, 1 ed.* Academic Press, 294p.

Garcia, J. A., Laos, D. L., Vega, N. V., Bendezú, M. D. R., Yarasca, P. E., Guilhermo, J. J., Surco-Laos, F. (2020). Actividad antioxidante y Antimicrobiana del extrato etanólico de las partes aéreas de *Solanum radicans* L. F. "hyallpachaqui" [Antioxidant and antimicrobial activity of the ethanolic extract of the aerial parts of Solanum radicans L. F. "hyallpachaqui"]. *Revista de la Sociedad Química del Perú* [*Journal of the Chemical Society of Peru*], 86(1), 5-12.

Ginzberg, E., Tokuhisa, J. G., Veilleux, R. E. (2009). potato steroidal glycoalkaloids: biosynthesis and genetic manipulation. *Potato Research*¸52, 1-15.

Gomes, A. D., Lima, R. A. (2014). Identificação de classes de metabólitos secundários no extrato etanólico dos frutos de *Solanum acanthodes* Hook e seu potencial fungicida sobre *Candida albicans* in vitro [Identification of classes of secondary metabolites in the ethanolic extract of Solanum acanthodes Hook fruits and their fungicidal potential against Candida albicans in vitro]. *Revista Eletrônica de Gestão, Educação e Tecnologia Ambiental* [*Electronic Journal of Environmental Management, Education and Technology*], 18(2), 736–744.

Guimarães, M. A. (2018). Avaliação terapêutica *in vivo* dos alcaloides epiisopiloturina e epiisopilosina extraídos de *Pilocarpus microphyllus* para esquistossomose. 125f. Tese. (Doutorado em Biotecnologia) – Programa de Pós-Graduação em Biotecnologia, Rede Nordeste de Biotecnologia, Universidade Federal do Piauí [In vivo therapeutic evaluation of epiisopiloturine and epiisopilosine alkaloids extracted from Pilocarpus microphyllus for schistosomiasis. 125f. Thesis. (*Doctorate in Biotechnology) – Graduate Program in Biotechnology, Northeast Biotechnology Network*, Federal University of Piauí].

Hassine, T., Mansour, A., Hammami, S. (2013). Case report of fatal poisoning by *Nicotiana tabacum* in Cartle in Tunisia. *Revue Médicine Véterinaire* [*Journal of Veterinary Medicine*], 164(3), 141-144.

Hernández, L. O., Carranzar, P. R., Cobos, L. E. P., López, L. I. L., Ascasio, J. A. V., Silva, S. Y. B. Bioguided fractionation From *Solanum elaeagnifolium* to Evaluate Toxicity on Cellular Lines and Breast Tumor Explants. *Revista de la Facultad de Ciencias Farmacêuticas y Alimentarias* [*Journal of the Faculty of Pharmaceutical and Food Sciences*], 24(2), 124-131.

Hernández-Herrera, R. M., Santacruz-Ruvalcaba, F., Ruiz-López, M. A., Norrie, J. (2014). Effect of liquid seaweed extracts on groeth of tomato seedllings (*Solanum lycopersicum* L.). *Journal of Applied Phycology*, 26(1), 619-628.

Hu, X. L., Gao, L. Y., Niu, Y. X., Tian, X., Wang, J., Meng, W. H., Zhao, Q. C. (2015). A neuroproteção por Kukoamine A contra o estresse oxidativo pode envolver receptores N-metil-d-aspartato [Neuroprotection by Kukoamine A against oxidative stress may involve N-methyl-d-aspartate receptors]. *Biochimica et Biophysica Acta - Assuntos Gerais*, 1850(2), 287-298.

Huang, W., Serra, O., Dastmalchi, K., Jin, L., Yang, L., Stark, R. (2017). Comprehensive MS and Solid-State NMR Metabolomic Profiling Reveals Molecular Variations in Native Periderms from Four Solanum tuberosum Potato Cultivars. Journal of *Agriultural and. Food Chemistry,* 65, 2258−2274.

Ikeda, T., Tsumagari, H., Honbu, T., & Nohara, T. (2003). Cytotoxic activity of steroidal glycosides from solanum plants. *Biological & Pharmaceutical Bulletin*, 26(8), 1198–1201.

Ito, S., Ihara, T., Tamura, H., Tanaka, S., Ikeda, S. Ikeda, T., Kajihara, H., Dissanayake, C., Abdel-Motaal, F. F., El-Sayed, M. (2007). α-Tomatine, the major saponin in tomato, induces programmed cell death mediated by reactive oxygen species in the fungal pathogen *Fusarium oxysporum*. *FEBS Letters*, 581(17), 3217-3222.

Jamwal, K., Bhattacharya, S., Puri, S. (2018). Plant growth regulator mediated consequences of secondary metabolites in medicinal plants. *Journal of Applied Research on Medicinal and Aromatic Plants,* 9, 26-38.

Jarald, E. E., Edwin, S., Saini, V., Deb, L., Gupta, V. B., Wate, S. P., & Busari, K. P. (2008). Anti-inflammatory and anthelmintic activities of Solanum khasianum Clarke. *Natural product research*, *22*(3), 269–274.

Jayakumar, K., Murungan, K. (2016). *Solanum* Alkaloids and their Pharmaceutical Roles: A Review. *Journal of Analytical & Pharmaceutical Research*, 36(6), 2-14.

Kasetti, R. B., Nabi, S. A., Swapna, S., Apparao, C. (2012). Cinnamic acid as one of the antidiabetic active principle(s) from the seeds of Syzygium alternifolium. *Food and Chemical Toxicology*, 50, 1425–1431.

Kaunda, J. S., Zhang, Y. J. (2019). The Genus *Solanum*: An Ethnopharmacological, Phytochemical and Biological Properties Review. *Natural Products and Bioprospecting,* 9(2), 77-137.

Kohara, A., Nakajima, C., Yoshida, S. and Muranaka, T. (2007). Characterization and engineering of glycosyltransferases responsible for steroid saponin biosynthesis in Solanaceous plants. *Phytochemistry*, 68, 478-486.

Kozukue, N., Friedman, M. (2003). Tomatine, chloropphill, β-carotene and lycopene content in tomatoes during growth and maturations. *Journal of the Science of Food and Agriculture,* 83, 195-200.

Lecanu, L., Hashim, A. I., Mccourty, A., Giscos-Douriez, I., Dinca, I., Yao, W., Vicini, S., Szabo, G., Erdélyi, F., Greeson, J., Papadopoulos, V. (2011). The naturally occurring steroid solasodine induces neurogenesis in vitro and in vivo. *Neuroscience*, 183, 251-264.

Leyton, V., Henderson, T. O., Mascara, D., Kawano, T. (2005). Atividade moluscicida de princípios ativos de folhas de *Lycopersicon esculentum* (Solanales, Solanaceae) em *Biomphalaria glabrata* (Gastropoda, Planorbidae) [Molluscicidal activity of active principles from leaves of Lycopersicon esculentum (Solanales, Solanaceae) on Biomphalaria glabrata (Gastropoda, Planorbidae)]. *Iheringia. Série Zoologia*, 95(2), 213-216.

Li, Y., Di, R., Hsu, W., Huang, Y., Cheung, H. (2015). Quality control of Lycium chinense and Lycium barbarum cortex (Digupi) by HPLC using kukoamines as markers. *Chinese Medicine (United Kingdom)*, 12(1). 2-8.

Lima, C. B. N. (2019). Extratos vegetais de Solanaceae: investigação química e bioatividade sobre *Sitophilus zeamais* Mots. (Coleoptera: Curculionidae) em grãos de milho armazenado. 84f. Dissertação. (Mestrado em Ciências/Entomologia) – Universidade de São Paulo [*Solanaceae plant extracts: chemical investigation and bioactivity on Sitophilus zeamais Mots.* (Coleoptera: Curculionidae) in stored corn grains. 84f. Dissertation. (Master of Science/Entomology) – University of São Paulo].

Lima, D. F., de Lima, L. I., Rocha, J. A., de Andrade, I. M., Grazina, L. G., Villa, C. (2017). Variação sazonal nos principais alcalóides de jaborandi (*Pilocarpus microphyllus*Stapf ex Wardleworth), uma espécie economicamente importante da flora brasileira [Seasonal variation in the main alkaloids of jaborandi (Pilocarpus microphyllusStapf ex Wardleworth), an economically important species of Brazilian flora]. *PLoS ONE* 12(2).

Lima, R. A., dos Santos, M. R. A., Smozinski, C. V. (2014). [Orgs.]. Flora de Rondônia, Brasil: *Solanum* L. (Solanaceae) [Flora of Rondônia, Brazil: Solanum L. (Solanaceae)]. Porto Velho-RO: *EDUFRO*, 102 p.

Liu, L., Hudgins, W. R., Shack, S., Yin, M. Q., Samid, D. (1995). Cinnamic acid: a natural product with potential use in cancer intervention. *International Journal of Cancer*, 62, 345-350.

Martins, G., Z., Moreira, R., R., D., Planeta, C., S., Almeida, A., E., Bastos, J., K., Salgueiro, L., Cavaleiro, C., Souza, M. C. (2015). Effects of the extract and glycoalkaloids os *Solanum lycocarpum* St. Hill on *Giardia lamblia* trophozoites, *Pharmacognosy Magazine,* 11(1), 161-165.

Marttins, E. A. S., Goneli, A. L. D., Cardoso, C. A. L., Almeida, G. L. de., Gonçalves, A. A., S., Silva, C. B. da., (2021). Pos-harvesting of *Solanum paniculatum* L. leaves. Part II: Antioxidante activity and chemical composition. *Revista Brasileira de Engenharia Agrícola e Ambiental* [*Brazilian Journal of Agricultural and Environmental Engineering*], 25(1), 17-22.

Mendes, I. C., Amado, P. A., Morais, M. G., Lima, L. A. R. dos S. (2020). Phytochemical screening and evaluation of phytotoxic activity of *Solanum lycocarpum* (Solanaceae) ripe fruit. *Acta Brasiliensis* [*Journal of Brazil*], 4(1), 45-48.

Milner, S. E., Brunton, N. P., Jones, P. W., O' Brien, N. M., Collins, S. G., Maguire, A. R. (2011). Bioactivities of glycoalkaloids and their aglycones from *Solanum* species. *Agricultural and Food Chemistry*, 59(8), 3454-3484.

Moloudizargari, M., Mikaili, P., Aghajanshakeri, S., Asghari, H. M., Shayegh, J. (2013). Pharmacological and therapeutic effects of Peganum harmala and its main alkaloids. *Pharmacognosy Reviews,* 7(14), 199-2012.

Moreira, C. B., Lima, S. S., Esquibel, M. A., Sato, A. (2010). Solasodine accumulation in regenerated plants of *Solanum torvum* Sw. *Revista Brasileira de Plantas Medicinais* [*Brazilian Journal of Medicinal Plants*], 12(1), 73-79.

Munari, C. C., De Oliveira, P. F., Lima, I. M. De S., Martins, S. De P. L., Da Costa, J. De C., Bastos, J. K., Tavares, D. C. (2012). Evaluation of cytotoxic, genotoxic and antigenotoxic potential of *Solanum lycocarpum* fruits glicoalkaloid extract in V79 cell. *Food and Chemical Toxicology,* 50(10), 3696-3701.

Murani, C., C., de Oliveira, P., F., Campos, J., C., L., Martins, S., P., L., Costa, J., C., Bastos, J., K., Tavares, D., C. (2014). Antiproliferative activity of *Solanum lycorcapum* alkaloidic extract and their constituents, solamargine and solasonine, in tumor cell lines, *Journal of Natural Medicines*, 68(1), 236-241.

Ngbolua, K. N., Mbadiko, C. M., Matondo, A., Bongo, G. N., Inkoto, C. L., Gbolo, B. Z., Lengbiye, E. M., Kilembe, J. T., Mwanangombo, D. T., Ngoyi, E. M., Falanga, C. M., Tshibangu, D. S. T., Tshilanda, D. D., Mpiana, P. T. (2020). Review on Ethno botany, Virucidal Activity, Phytochemistry and Toxicology of *Solanum genus*: Potential Bio-resources for the Therapeutic Management of Covid-19. *European Journal of Nutrition & Food Safety*, 12(7), 35-48.

Niño, J., Correa, Y. M., Mosquera, O. M. (2009). Biological activities of steroidal alkaloids isolated from *Solanum leucocarpum*. *Pharmaceutical Biology*, 47(3), 255-259.

Oliveira, M. L. B., França, T. A. T., Cavalcante, F. S., Lima, R. A. (2020). Uso, classificação e diversidade de *Solanum* L. (Solanaceae) [Use, classification and diversity of Solanum L. (Solanaceae)]. *Biodiversidade* [*Biodiversity*], 19(3), 142-159.

Olmstead, R. G. (2016). Phylogeny and biogeography in Solanaceae, Verbenaceae and Bignoniaceae: a comparison of continental and intercontinental diversification patterns. *Botanical Journal of the Linnean Society*, 171(1), 80-102.

Palomino K. (2014). Efecto sinérgico in vitro entre el extracto etanólico de *Solanum sessiliflorum* "cocona" y vancomicina sobre el crecimiento de *Staphylococcus aureus* resistente a meticilina. (Tesis para optar el Grado de Bachiller en Medicina). Trujillo: Universidad Nacional de Trujillo [Synergistic effect in vitro between the ethanolic extract of Solanum sessiliflorum "cocona" and vancomycin on the growth of Staphylococcus aureus resistant to methicillin. (*Thesis to choose the Degree of Bachelor of Medicine*). Trujillo: National University of Trujillo].

Parr, A. J., Mellon, F. A., Colquhoun, I. J., Davies, H. D. (2005). Dihydrocaffeoyl polyamines (kukoamine and allies) in potato (*Solanum tuberosum*) tubers detected during metabolite profiling. *Journal of Agricultural and Food Chemistry*, 53, 5461−5466.

Pereira, A. P. A., Angolini, C. F. F., Adani, H. B., Usberti, F. C. S., Paulino, B. N., Clerici, M. T. P., Neri-Numa, I. A., Moro, T. de M. A., Eberlin, M. N., Pastore, G. M. (2021). Impact of ripening on the health-promoting components fruta-do-lobo (*Solanum lycocarpum* St. Hill). *Food Research International*, 139, 110589.

Pereira, A. P. A., Angolini, C. F. F., De Souza-Sporkens, J. C., Da Silva, T. A., De Oliveira, H. C. F., Pastore, G. M. (2021) Brazilian sunberry (*Solanum oocarpum* Sendtn): Alkaloid composition and improvement of mitochondrial functionality and insulin secretion of INS-1E cells. *Food Research International*, 1, 110589.

Pereira, A. P., Angolini, C. F., Paulino, B. N., Lauretti, L. B., Orlando, E. A., Silva, J. G., Neri-Numa, I. A., Souza, J. D., Pallone, J. A., Eberlin, M. N., Pastore, G. M. (2019). A comprehensive characterization of Solanum lycocarpum St. Hill and Solanum oocarpum Sendtn: Chemical composition and antioxidant properties. *Food Research International,* 124, 61–69.

Pereira, I. S., Rodrigues, V. F., Vega, M. R. (2016). Flavonoides do Gênero *Solanum* [Flavonoids of the Solanum genus]. *Revista Virtual de Química* [*Virtual Chemistry Magazine*], 8(1), 4-26.

Pinto, F. D. C. L., Uchoa, D. E. D. A., Silveira, E. R., Pessoa, O. D. L., Braz-Filho, R., Silva, F. M., Espindola, L. S. (2011). Glicoalcaloides antifúngicos, flavonoides e outros constituintes químicos de *Solanum asperum* [Antifungal glycoalkaloids, flavonoids and other chemical constituents of Solanum asperum], *Química Nova* [*New Chemistry*], 34(2), 284-288.

Ramírez, A., Isaza, G., Perez, J., Martinez, M. (2017). Estudio fitoquímico preliminar y evaluación de la actividad antibacteriana de *Solanum dolichosepalum* bitter (frutillo) [Preliminary phytochemical study and evaluation of the antibacterial activity of Solanum dolichosepalum bitter (frutillo)]. *Revista Cubana de Plantas Medicinais* [*Cuban Journal of Medicinal Plants*], 18(1), 21-25.

Ren, Y., Shen, L., Zhang, D. W., Dai, S. J. (2009). Two new sesquiterpenoids from *Solanum lyratum* with cytotoxic activities. *Chemical and Pharmaceutical Bulletin,* 57(4), 408-410, 2009.

Sampaio, V. S., Vieira, I. M. F., Lima-Júnior, E. A., Loiola, M. I. B. (2019). Flora do Ceará, Brasil: *Solanum* (Solanaceae) [Flora of Ceará, Brazil: Solanum (Solanaceae)]. *Rodriguésia*, 70(2), 12-17.

Sampaio, V. S., Araújo, N. D., Agra, M. de F., (2014). Characters of leaf epidermis in Solanum (clade Brevantherum) species from Atlantic Forest of Northeastern Brazil. *South African Journal of Botany,* 94, 108-113.

Samuels J. (2015). Biodiversity of food species of the Solanaceae family: a preliminary taxonomic inventory of subfamily Solanoideae. *Resources,* 4(2), 277-322.

Silva, T. M. S., Camara, C. A., Freire, K. R. L., Silva, T. G., Agra, M. F., Bhattacharyya, J. (2008). Steroidal glycoalkaloids and molluscicidal activity of Solanum asperum Rich. Fruits. *Journal of the Brazilian Chemical Society,* 19, 1048.

Simões, C. M. O. (2017). Farmacognosia do produto natural ao medicamento. 1 ed. Porto Alegre: *Artmed* [*Pharmacognosy from the natural product to the medicine. 1 ed. Porto Alegre: Artmed*], 502 p.

Simões, C. M. O., Schenkel, E. P., Gosmann, G., Mello, J. C. P., Mentz, L. A., Petrovick, P. R. (1999). *Farmacognosia: da planta ao medicamento.* 1°. ed., Ed. Universidade -

UFRGS/Ed. da UFSC: Porto Alegre/Florianópolis [*Pharmacognosy: from plant to medicine. 1st. ed.*, Ed. University - UFRGS/Ed. from UFSC: Porto Alegre/Florianópolis].

Simões, C. M. O., Schenkil, E. P., Gosmann, G., De Mello, J. C. P., Mentz, L. A., Petrovick, P. R. (2002) In: *Farmacognosia: da planta ao medicamento.* 4º ed. Porto Alegre-RG. *Ed. UFRGS, cap 29, Alcalódes: Generalodades e Aspctos Básicos* [*Pharmacognosy: from plant to medicine.* 4th ed. Porto Alegre-RG. Ed. UFRGS, cap 29, Alkalodes: Generalities and Basic Aspects], 651-666p.

Sinani, S. S. S. A., Eltayeb, E. A. (2017). The steroidal glycoalkaloids solamargine and solasonine in *Solanum* plants. *South African Jornal of Botany*, 112, 253-269.

Soares-Mota, M. R., Schwarz, A., Bernardi, M. M., Maiorka, P. C., Spinosa, H. S. (2010). Toxicology evaluation of 10% *Solanum lycocarpum* St. Hill fruit consumption in the diet of growing rats: hematological, biochemical and histopathological effects. *Experimental and Toxicologic Pathology*, 62, 549–553.

Souto, L. F. L., Da Silva, A. J. R. (2020). Calistegines, polyhydroxylated alkaloids that inhibit glycosidases presente in *Solanum crinitum. Brazilian Journal of Development*, 6(8), 55202-55215.

Stehmann, J. R., Mentz, L. A., Agra, M. F., Vigoli-Silva, M., Giacomin, L., Rodrigues, I. M. C. (2015). [Internet]. Solanaceae in Lista de Espécies da Flora do Brasil. *Jardim Botânico do Rio de Janeiro* [*Solanaceae in List of Brazilian Flora Species. Botanical Garden of Rio de Janeiro*]. Available in: http://floradobrasil.jbrj.gov.br/jabot/floradobrasil/FB14716.

Sun, L. X., Fu, W. W., Ren, J., Xu, L., Bi, K. S., Wang, M. W. (2006). Cytotoxic constituents from *Solanum lyratum. Archives of Pharmacal Research,* 29(2), 135-139, 2006.

Suthar, A. C., Mulani, R. M. (2008). A high-performance thin layer chromatography method for quantitative estimation of Diosgenin in *Solanum nigrum. Pharmacognosy Magazine*, 4(14), 112-115.

Vaz, N. P., Costa, E. V., Santos, E. L., Mikich, S. B., Marques, F. A., Braga, R. M., Delamelina, C., Duarte, M. C. T., Ruiz, A. L. T. G., Souza, V. H. S., Carvalho, J. E. de., Maia, B. I. N. (2012). Caavuranamide, a novel steroidal alkaloid from the ripe fruits of *Solanum caavurana* Vell. (Solanaceae). *Journal of the Brazilian Chemical Society,* 23(2), 361-366.

Vieira, P. M., Marinho, L. P. M., Ferri, S. C. S., Chen-Chen, L. (2013). Protective effects of steroidal alkaloids isolated from *Solanum paniculatum* L. against mitomycin cytotoxic and genotoxic actions. *Anais da Academia Brasileira de Ciências* [*Proceedings of the Brazilian Academy of Sciences*], 85(2), 553-560.

Vizzotto, M.; Krolow, A. C.; Weber, G. E. B. (2010). Metabólitos Secundários Encontrados em Plantas e sua Importância. *Embrapa Clima Temperado*. Embrapa: Pelotas [*Secondary Metabolites Found in Plants and Their Importance. Embrapa Temperate Climate*. Embrapa: Pelotas], 16p.

Walker, B. W., Manhanke, N., Stine, K. J. (2008). Comparison of the interaction of tomatine with mixed monolayers containing phospholipid, egg sphingomyelin and sterols. *Biochem., Biophys., Acta.,* 1778, 2244-2257.

Wang, Y., Wang, S. L., Zhang, J. Y., Song, X. N., Zhang, Z. Y., Li, J. F., Li, S. (2018). Anti-ulcer and anti-Helicobacter pylori potentials of the ethyl acetate fraction os *Physalis alkekengi* L. var. franchetii (Solanaceae) in rodent. *Jornal of Ethnopharmacology*, 211(30), 197-206.

Weissenberg, M. (2001). Isolation of solasodine and other steroidal alkaloids and sapogenins by direct hydrolysis-extraction of *Solanum* plants or glycosides therefrom. *Phytochemistry,* 58(3), 501-508.

Wink, M. (2003). Evolution of secondary metabolites from an ecological and molecular phylogenetic perspective. *Phytochemistry*, 64(1), 3-19.

Wu, S., Meyer, R., Whitaker, B., Litt, A., Kenelly, E. A new liquid chromatography–mass spectrometry-based strategy tointegrate chemistry, morphology, and evolution of eggplant (*Solanum*) species. Wu, S.-B., Meyer, R. S., Whitaker, B. D., Litt, A., & Kennelly, E. J. (2013). A new liquid chromatography–mass spectrometry-based strategy to integrate chemistry, morphology, and evolution of eggplant (*Solanum*) species. *Journal of Chromatography A*, 1314(1), 154–172.

Wu, T., Haiyang, N., Fengzhong, W., Yi, W. (2016). Characterization of Polyphenols from Lycium ruthenicum Fruit by UPLC-Q-TOF/MSE and Their Antioxidant Activity in Caco-2 Cells. J. *Agriultural and. Food Chemistry,* 64(11), 2280-2288.

Yoshikawa, M., Nakamura, S., Ozaki, K., Kumahara, A., Morikawa, T., & Matsuda, H. (2007). Structures of Steroidal Alkaloid Oligoglycosides, Robeneosides A and B, and Antidiabetogenic Constituents from the Brazilian Medicinal Plant *Solanum lycocarpum*. *Journal of Natural Products*, 70(2), 210–214.

Zhang, D. W., Li, G. H., Yu, Q. Y., Dai, S. J. (2010). New anti-inflammatory 4-hydroxyisoflavans from *Solanum lyratum*. *Chemical and Pharmaceutical Bulletin,* 58(6), 840-842, 2010.

Chapter 5

Catharanthus roseus L. (Apocynaceae): An Alkaloid-Rich Herb with Diverse Pharmacological Activities

Arjun Pandian[1,*], Raju Ramasubbu[2], Kaliyaperumal Ashokkumar[3] and Ruchi Badoni Semwal[4]

[1]Division of Research and Innovation, Department of Biotechnology, Saveetha School of Engineering, Saveetha Institute of Medical and Technical Sciences, Thandalam, Chennai, Tamil Nadu, India
[2]Department of Biology, The Gandhigram Rural Institute (Deemed to be University), Gandhigram, Dindugal, Tamil Nadu, India
[3]School of Agriculture and Animal Sciences, The Gandhigram Rural Institute (Deemed to be University), Gandhigram, Dindigul, Tamil Nadu, India
[4]VSKC Government Postgraduate College, Dakpathar, Dehradun, Uttarakhand, India

Abstract

Vinblastine, vincristine, vinleurosine, vindoline, and catharanthine are notable monoterpenoid indole alkaloids consequential from the plant periwinkle. Vinblastine, vincristine, vinleurosine, and catharanthine are anticancer compounds. *Catharanthus roseus* belongs to the family of Apocynaceae and has pink and white flowered, used numerous illnesses, including diabetes, cancer, high blood pressure (HBP), constipation, asthma, menorrhagia, rheumatism, indigestion, skin conditions, bloody diarrhoea, and sadness are treated with it. Catharanthine is approximately completely accumulated in the wax exudates on the plant leaf surface, while vindoline is fashioned in specialized leaf cells internally. The plant

---

* Corresponding Author's Email: arjungri@gmail.com.

In: The Essential Guide to Alkaloids
Editor: Deepak Kumar Semwal
ISBN: 979-8-88697-456-0

has numerous pharmacological effects, including antioxidant, antimicrobial, antifeedant, anti-sterility, anthelmintic, antidiabetic and antidiarrheal. Due to independent enlargement in liquid medium exclusive of the use of exogenous hormones, established biochemical and genetic expression, and the potential for terpenoid indole alkaloids such as serpentine, ajmalicine, phenylalanine, tyrosine, tabersonine, catharanthine, and lochnericine to be elicited by methyl jasmonate, chitosan, and fungal elicitation, hairy root culture may perhaps entirely scale up to an industrial scale.

Keywords: antidiabetic, ajmalicine, cancer, lochnericine, vinblastine, vincristine

1. Introduction

The Apocynaceae family includes the tropical perennial medicinal plant *Catharanthus roseus* L. It is a herb that cultivates undomesticated in subtropical and tropical regions of humankind and is extensively sophisticated. It is native to and also endemic to Madagascar. It is a pantropical species that is mostly found in the West Indies and Madagascar. There are two popular varieties of *C. roseus* called "Rosea" and "Alba," which have pink flowers and white flowers, respectively (Das and Sharangi, 2017). *C. roseus* can be multiplied through seeds or cuttings (Aslam et al., 2010). The blooms have five-lobed petals and range in colour from white to shadowy pink with a dark crimson core. The flowers have a base tube diameter of roughly 2.5–3 cm (Kabesh et al., 2015). The wild plant's blossoms are normally light pink with a violet centre, but gardeners have created more than 100 variations with hues ranging from white to pink to purple (Aslam et al., 2010). More than 200 important secondary metabolites are recognized to be fashioned through the plant (chiefly alkaloids). Alkaloids are found in vinca, primarily vincristine and vinblastine (Das and Sharangi, 2017).

The *Coriaceus, Lanceus, Ovalis, Pusillus, Scitulus, Trychophyllus,* and *Longifolius* species of Clostridium) (Gupta et al., 2007). This plant, also referred to as vinca in the trade, is a pantropical species that is currently found all over the world. Cultivated as a garden plant in Iraq and called "cat eyes" for its rose and white colouring, typically terpene indole alkaloids are acknowledged to be shaped through the plant, counting the anticancer drugs are vincristine and vinblastine, which are fashioned in diminutive amounts and based on the plant's developmental and physiological stages. However, in most cases, these compounds have not been commercially viable. Less than 1% of the world's carbon is typically produced in very small amounts of these chemicals (Cragg and Newman, 2005). To produce 1 g and 20 mg respectively, 1000 kg of leaves and stems are needed. It has been occasion experienced and supported through people's beliefs that *C. roseus* works as a folk medicine in numerous nations approximately the humanity for a variety of ailments, including diabetes, high blood pressure (BP), cancer, constipation, asthma, menorrhagia, indigestion, rheumatism, skin conditions and depression (Kabesh et al., 2015). All parts of the vinca plant, including the leaves, roots, flowers, and stems, are employed extensively in contemporary medicine as potential anticancer agents. Other significant alkaloids that are progenitors to the vinblastine and vincristine group of alkaloids include catharanthine (Kabesh et al., 2015) and vindoline (Gupta et al., 2005; Aslam et al., 2010).

The medicinal plant *C. roseus* has an extensive assortment of pharmacological actions, including antibacterial, anthelmintic, antifeedant, anti-sterility, antioxidant, anti-diarrheal, and antidiabetic properties (Kabesh et al., 2015). To make use of the higher herbal value of this plant, more research will be put into *C. roseus*. Due to its chemical makeup, which includes more than 130 dissimilar terpenoid indole alkaloids (TIAs), including a quantity that is potent and pharmacologically important, such as vinblastine and vincristine, *C. roseus* has concerned a lot of pharmaceutical concentration. These are demonstrated through the extremely successful diabetic, anticancer, antioxidant and hypertension treatments (Van Der et al., 2004). Two well-known *C. roseus* TIAs known as the first all-natural treatments and preventatives for malignant tumour disorders are vinblastine and vincristine (Sottomayor and Barcelo, 2006). Because the total plant concentrations of vincristine and vinblastine are so low (about 0.0005% w/w) (Van Der et al., 2004), the extraction of *C. roseus* is exceedingly expensive, it is predictable that 500 kg of dried *C. roseus* leaves makes available merely 1 g of vinblastine. Alkaloids are among the majority frequent classes of derived metabolites found in plants. They are glycosides and are highly prized for their potent anticancer properties. Alkaloids can be separated into three most important categories. The original grouping includes indole alkaloids like vincristine and vinblastine from *C. roseus*. The second group consists of quinoline alkaloids like *Camptotheca acuminate* produced camptothecin, while the final collection consists of diterpenoid alkaloids like taxol, which are comparable to *Taxus* and *Corylus* species (Sahi et al., 2022).

It is extremely imperative that the consumption of medicinal plants in conventional medicine was established to be successful, contemptible and meticulous (Van Der et al., 2004). In very little amounts, the plant's aerial portions produce these important alkaloids. Researchers are using the plant tissue culture performance to boost the alkaloid content of *C. roseus in vitro* due to the high expense of drug isolation. Researchers have been attempting to improve the fabrication of the important antineoplastic alkaloids vincristine and vinblastine for a long time. Nevertheless, the TIAs pathways are not solitary restricted through peripheral abiotic and biotic stimulus, but also through tissues predominantly and throughout improvement.

The elevated level of specialisation of several leaf cells where the congregation of the anticancer compound vincristine occurs contrasts with the low levels of the anticancer compounds vincristine and vinblastine establish in plants, constructing their withdrawal and exceptionally luxurious development. These squat levels are largely interrelated to the spatial disconnection of biosynthetic sites someplace these compounds are produced in the plant and to an elevated degree (Van Der et al., 2004). In reality, vindoline is formed in particular interior leaf cells while catharanthine gathers approximately exclusively in wax exudates on the surface of the leaf, indicating that transport activities must be involved for their combination to happen (Sottomayor et al., 2006). Freshly, CrTPT2, an ABC transporter whose main job is to improve transport and, as a result, the buildup of catharanthine in leaf epidermal surface, was discovered (Van Der et al., 2004).

Whereas AVLB was essentially exposed to be deficient in extremely youthful leaves (Naaranlahti et al., 1991), the catharanthine and vindoline corporeal disconnection experimental through Yu and DeLuca (2013) is in all probability a preventive feature in urbanized leaves, where the dimer AVLB was frequently reported to be plentiful (Carqueijeiro et al., 2013). In information, using vacuoles isolated from established plant leaves, it was shown that vindoline, catharanthine and AVLB were vacuoles of mesophyll cells accumulated through an exceptional proton antiport system, dependent on the transtonoplast pH ascent fashioned through V-H+-PPase and V-H+-ATPase. Significant monoterpenoid indole alkaloids obtained from the plant periwinkle, *C. roseus* about 130 alkaloids, comprise vincristine (VCR), vinblastine (VBL), vindoline (VDL), catharanthine (CTR) and vinleurosine (VLS) (Noble 1990).

Currently, Hodgkin's ailment and neoplasms such bladder, breast, lymphomas, lung, and leukaemia are treated with a lot of VBL and VCR (Sertel et al., 2011). They are referred to be antimitotic, antimicrotubule, or mitotic inhibitors because they prevent the distribution and proliferation of cancer cells (Okouneva et al., 2003). Hodgkin's sickness heightened lymphoblastic leukaemia, and lymphoblastosarcoma are all treated with VLS. Precursors for the synthesis of VBL include CTR and VDL (Pennanen and Huhtikangas, 1990). Due to the outline levels of VBL and VCR in the plant (Mu et al., 2012), it has been challenging to quantitatively analyse the structures of VCR, VBL, VDL, VLS and CTR, numerous investigate have been described in an effort to enhance the fortitude of these alkaloids in *C. roseus* (Tikhomiroff and Jolicoeur, 2002). HPLC with fluorescence, UV absorbance, or electrochemical detection is the major technique now used to determine these alkaloids (Horvath and Ivanyi 1982). The quantification and analysis of *Catharanthus* alkaloids also employ high-performance thin layer chromatography (Horvath and Ivanyi, 1982), gas chromatography, supercritical fluid chromatography, radioimmunoassay (Deus-Neumann et al., 1987), and capillary zone electrophoresis (Zhou et al., 1993; Lin et al., 2014). This worldwide cultivated plant yields terpenoid indole alkaloids (TIAs) that are used in the pharmaceutical sector. The *C. roseus* leaves are used to extract the dimeric indole alkaloids vincristine and vinblastine, which are both powerful cancer treatments. Vinblastine and vincristine bind to tubulin in cell biology, causing polymerization that prevents the synthesis of microtubules. Vinblastine and vincristine's microtubule poisoning causes the mitotic spindle to be disrupted, which ultimately causes metaphase to be arrested (Zandi, 2021). This inhibition also causes the spindle assembly checkpoint proteins to be activated. It then stops the cell cycle from continuing in order to allow for spindle realignment and stop abnormal chromosomal separation. The dysfunctional mitotic congregation checkpoint proteins and incorrect chromosomal isolation resulted in a mitotic catastrophe, cell assassination machinery mediated through abnormal mitosis (Bagnyukova et al., 2010).

The main drawback of cell cultures of these drugs is their yield is very low, pharmacologists and chemists are particularly interested in humanizing the production of catharanthine in *C. roseus* cell cultures since vindoline and catharanthine can be combined to appearance vinblastine in elevated yields, and vindoline is widely available in plants. Elicitors can also control how much of these alkaloids are produced. Due to stumpy vincristine and vinblastine contents in the plants, metabolic engineering for alternative production methods was promoted and made practicable (Aslam et al., 2002). It is useful to combine commercially available catharanthine and vindoline using techniques in engineering like semi synthesis, complete even chemical or chemical synthesis (Tam et al., 2010) or enzymatic coupling (Kuboyama et al., 2004).

Due to the chemical makeup of *C. roseus*, which contains more than 130 dissimilar indole terpenoid alkaloids (TIAs), several of which are potent and important pharmacologically, such as vinblastine and vincristine, which have been shown to be highly successful in the management of cancer, hypertension, and diabetes. Two well-known *C. roseus* TIAs, vinblastine and vincristine, are regarded as the first all-natural treatments for malignant tumour disorders. Vincristine and vinblastine are both exceedingly expensive due to the high cost of *C. roseus* extraction and the accumulated concentrations from the entire plant. The geographical separation of the plant's biosynthetic sites, including the highly specialised TIAs production route in leaf cells, is largely responsible for the low amounts of vincristine and vinblastine. Scientists are particularly interested in finding ways to augment vincristine and vinblastine production by utilizing a variety of biological techniques because of the low recovery of these drugs from the expensive extraction method. The chemical structures of vincristine, vinblastine, vinorelbine and vinflunine are given in Figure 1.

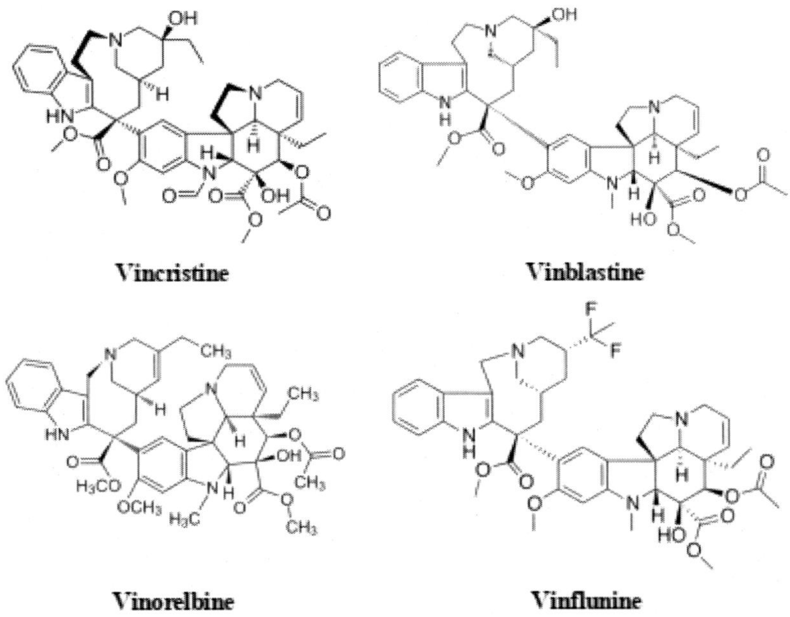

Figure 1. Chemical structures of key bioactive alkaloids from *Catharanthus roseus*.

Table 1. Strategy to augment the production of terpenoid indole alkaloids (TIAs) in *C. roseus*

S.No	Strategy	Materials	Observations	Total TIAs	Ref.
1	Medium with mannitol (250 mM)	Compact callus cluster	5.05 Catharanthine increased (19.7 mg/L) 2.77 Serpentine increased (13.6 mg/L) 4.54 Ajmalicine increased (42.3 mg/L)	4.13 increase (75 mg/L)	Zhao et al., 2001
2	Medium with Sucrose (50 g/L)	Compact callus cluster	1.25 Serpentine increased (10 mg/L) 1.20 Ajmalicine increased (14.5 mg/L)	1.16 increase (46.7 mg/L)	Zhao et al., 2001
3	Cell density high (200 g FW/L)	Suspension cultured cells	1.20 Ajmalicine increased (60 mg/L)	ND	Almagro et al., 2011
4	Supply with succinic acid (10 mM)	Compact callus cluster	3.5 Catharanthine increased (7 mg/L) 7.5 Serpentine increased (15 mg/L) 16 Serpentine increased (32 mg/L)	4.86 increase (73 mg/L)	Zhao et al., 2001
5	Elicitation with Na Nitroprusside (0.1 mM)	Hairy root culture	2 Lochnericine increased (1 mg/g DW) 2.3 Tabersonine increased (0.7 mg/g DW) 2 Ajmalicine increased (0.7 mg/g DW)	1.42 increase (3.7 mg/g DW)	Li et al., 2011
6	Elicitor with MeJA (101.9 µM)	Suspension cultured cells	2 Tabersonine increased (3.8 mg/g DW)	1.33 increase (2.2 mg/g DW)	Goldhaber et al., 2014
7	Elicitor with MeJA (100 µM)	Suspension cultured cells	27.44 Ajmalicine increased (137.2 mg/L) 11.12 Catharanthine increased (55.6 mg/L)	ND	Almagro et al., 2014
8	Elicitation with MeJA (250 µM)	Hairy root culture	7 Ajmalicine increased (6.34 mg/g DW) 2.9 Serpentine increased (1.71 mg/g DW) 3 Ajmaline increased (12 mg/g DW) 3 Catharanthine increased (4.34 mg/g DW)	1.32-fold increase (49 mg/L)	Ruiz-May et al., 2009

Table 1. (Continued)

S.No	Strategy	Materials	Observations	Total TIAs	Ref.
9	Elicitation with *Aspergillum niger* mycelium and tetramethyl ammonium bromide	Suspension cultured cells	21 Ajmalicine increased (63 mg/L) 17 Catharanthine increased (17 mg/L)	3.84 increase (96 mg/L)	Zhao et al., 2001
10	CrPrx Overexpression	Hairy root culture	5 Serpentine increased (3.7 mg/g DW) 3 Ajmalicine increased (0.35 mg/g DW)	1.5 increase (85 mg/g DW)	Jaggi et al., 2011
11	G10H/DXS Overexpression	Hairy root culture	1.35 tabersonine increased (0.9 mg/g DW) 1.15 Lochnericine increased (1.4 mg/g DW)	0.0072 mg/g DW	Peebles et al., 2011
12	Transient overexpression of transcription factor CrMPK3	Leaves	3.52 Serpentine increased (0.061 mg/g DW) 2.66 vindoline increased (4.1 mg/g DW) 1.44 Catharanthine increased (1.3 mg/g DW) 2 Vincristine increased (1.75 mg/g DW)	ND	Raina et al., 2012
13	TDC/STR Overexpression	*Morinda citrifolia* cell cultures	Strictosidine enhancement (21.2 mg/L)	ND	Hallard et al., 1997

By using *in vitro* culture, biotechnology has been successful in generating numerous varieties of therapeutic plants with elevated commotion. Hairy root culture is nearly everyone shows a potential biotechnological method for bulky amount synthesis of these substances. *Agrobacterium rhizogenes*, a soil bacterium, and plants interact to produce "hairy root," a valuable byproduct of the transformation process that can grow indefinitely in culture conditions (Hallard et al., 1997). Hairy roots may be thought of as a plant industrial unit that quickly produced comparable resultant metabolites and even advanced levels than the parents. Outstanding autonomous expansion in fluid medium lacking exogenous hormones, balanced biochemical and genetic expression, and hairy root culture's probable for occupied industrial balance extension (Hallard et al., 1997).

Precursors and elicitors have been suggested as a capable approach to increase the content and shorten the harvest period of expensive less important metabolites in hairy roots. Numerous prior studies have shown that precursor (tyrosine, phenylalanine) and elicitors (fungal, chitosan, methyl jasmonate elicitation) may accelerate the TIAs stuffing in CHR, such as serpentine, ajmal, ajmalicine, in addition, vinblastine and vincristine accelerations received less research, maybe as a result of their placement near the end of the bio-synthesis routes for TIAs (Raina et al., 2012). Different strategies to augment the production of terpenoid indole alkaloids (TIAs) in *C. roseus* are given in Table 1.

2. Clinical Pharmacology

2.1. Vincristine

Require of responsive assays for determining vincristine plasma concentration restricts pharmacokinetic investigations of the drug; however, plasma authorization is quick due to substantial tissue compulsory and extensive quantity of allocation. Vincristine is believed to have triexponential pharmacokinetics, with fast allocation after bolus management, segment allocation (50–155 min), and an abolition partially life of roughly 85 h. (Sethi and Kimball, 1981). Vincristine is developed through cytochrome (P450) 3A in the liver and concomitantly administer medications might moreover competitively hinder or encourage cytochrome (P450) 3A authorization of vincristine. Children materialize to have more advanced plasma authorization than adults (Zhou et al., 1993). Vincristine builds up in a variety of tissues, including the spleen, liver, pancreas, kidney, lung and bone marrow. It is largely not found in the eye, brain or fat tissue. Vincristine is eliminated in the bile and faeces as either the unmodified medication or a metabolite (Jackson, 1978). Renal excretion of vincristine is not very high. The association connecting the antitumor and plasma pharmacokinetics possessions of vincristine is not wholly defined.

Adults typically receive vincristine as a bolus intravenous combination (1.4 mg/m^2) (with the greatest amount of 2 mg), and children typically receive it at 1.5–2.0 mg/m2. Vincristine can be administered up to once per week, but the dosing regimen varies based on the category of cancer, the patient's response, and any other medications being taken concurrently. Neurotoxicity is the measure of preventive toxicity and the challenge to use an enduring combination as a substitute for bolus vaccination to condense neurotoxicity has been unconvincing. The vincristine dose is reduced in the surroundings of relentless severe neurotoxicity or hepatic dysfunction. Characteristically patients will entertain (50%) the considered dose for restrained hyperbilirubinemia, and relentless hyperbilirubinemia (25%) (Van den Berg, 1982).

2.2. Vinblastine

Vinblastine's triexponential pharmacokinetic model, which also applies to vincristine, is consistent with the pharmacokinetic revision of patients receiving bolus injections of the drug. Vinblastine leaves the plasma quickly and is transported to tissues, principally the liver, kidneys, spleen, and lung. Vinblastine's eradication half-life is around 29 h, and at 48 h, very little of the medication is still present in the body (Owellen and Hartke, 1975). Comparable to the supplementary alkaloids vinca, metabolized through vinblastine is the hepatic cytochrome (P450) 3A enzyme. Other drugs that either promote or inhibit the movement of cytochrome P450 3A may have an impact on this pathway and may affect patients with hepatic dysfunction. With limited renal excretion, vinblastine is primarily eliminated in the bile and faeces. Vinblastine is frequently administered at a measure (6 mg/m^2) to both children and adults, with adjustments made for hepatic dysfunction and haematological tolerance (Owellen et al., 1977).

2.3. Modes of Action of Vinblastine and Vincristine

The microtubules disruption, protein inhibition and nucleic acid synthesis, oxidised glutathione elevation, lipid metabolism alteration and membrane lipid contented, the elevation of cyclic adenosine monophosphate (cAMP), and awkwardness of calcium calmodulin synchronized cAMP phosphodiesterase are just a few of the numerous biochemical effects observed after exposure of cells and tissues to vinca alkaloids. The only well-documented direct effect of the vinca alkaloids is the microtubule disruption, which results from their reversible obligatory to tubulin. The vinca alkaloids are comparatively hydrophobic molecules that separate into lipid bilayers in the uncharged position, changing the function and structure of membranes (the subunit protein of microtubules). The majority of vinca alkaloids' concentrations have pharmacological activity, and the biochemical effects of exposure to vinca alkaloids are likely secondary to microtubule disruption, though it is potential that drug-induced changes in lipid bilayers may concern some membrane-independent processes. These substances cause the production of sizable crystalline aggregates made of tubulin and drugs at high intracellular concentrations (Kumar, 2016).

Vinca alkaloids have a variety of biochemical effects, but their anticancer activity is typically accredited to their capability to microtubule disrupt, which results in the termination of mitotic spindles and metaphase capture in separating cells. In addition to mitosis, microtubules are implicated in numerous other cellular processes, and introduction to vinca alkaloids results in a variety of physiologic consequences, many of which have the potential to impair vital functions in both non-dividing and dividing cells. Vinblastine and vincristine therapy has been associated with morphological alterations and cell death in non-dividing customary and leukemic lymphocytes, as well as in cultured leukemic cells in the interphase, G1 and S phases. Vinca alkaloids inhibit chemotaxis in human monocytes and directional movement of tumour cells in culture (Kumar, 2016). Microtubules are necessary for the movement of organelles along neural processes, including secretory granules and mitochondria, as well as the transportation of different metabolites. Vinca alkaloids cause neurotoxicity by preventing axonal transmission in nerve tissue. The vinca alkaloids appear to interfere with membrane trafficking and disturb the cytoskeleton to suppress secretory activities. After receiving therapy with vinca alkaloids, platelets, which rely on the stability of the peripheral ring of microtubules for their discoidal structure, develop a spherical shape. These few instances show how vinca alkaloids have an assortment of potentially cytotoxic effects that are unconnected to mitotic inhibition (Kumar, 2016).

Even though the belongings of vinca alkaloids on the structure and microtubule functions have been thoroughly studied, methodological issues have made it challenging to determine the environment and numeral of vinca alkaloid obligatory sites on tubulin. Nevertheless, it appears that every heterodimer of tubulin has one "vincaspecific" site with an elevated essential empathy and an undetermined number of nonspecific sites with stumpy sympathy. The discrepancies in test settings and techniques of ligand binding data interpretation make it difficult to contrast the tubulin-binding capacities of various vinca alkaloids. However, some generalisations are possible (Kumar, 2016).

The bioactive metabolites 4000 fungal derivations have been described and fungi associated with plants are sometimes able to produce identical bioactive compounds as the host plant itself. Fungi are one of the major sources of naturally occurring bioactive molecules. Following the identification of gibberellins from *Fusarium fujikuroi*, it has been hypothesised that plant and fungal intergeneric genetic exchange may have occurred. It was discovered and confirmed that endophytic fungi connected to *Taxus brevifolia* may produce Taxol, which opened up a fresh possibility. Because of this, we were inspired to isolate endophytic fungi from *C. roseus* in order to produce the highly valuable and low-volume drugs vincristine and vinblastine. The fungal source would lower the cost of the aforementioned life-saving medications and, in some cases, prevent the extinction of the plant (Kumar, 2016).

In addition to being a significant elicitor and being found to be high in yeast extract (YE), the vitamin B complex also contains vital substances like N-acetyl-glucosamine oligomers, chitin, glucan, ergosterol and glycopeptides. These substances cause the synthesis of metabolites, which in turn triggers plant protection responses. When the chemical structure of the active molecule is not fully understood, complicated biological preparations have also been used as elicitors in other cultural experiments. A number of the examined plant genera showed the excessive synthesis of an important phytocompound when YE was added to the culture. Stress is created in growing tissues as a result of the addition of exogenous substances, which enhances the manufacture of alkaloids. In order to assess the effect of YE on VC and VB yield on cultivated tissues/plants derived from protoplasts, *C. roseus* studied the activity of antioxidant enzymes and other pressure markers in cultured tissues that more than manufacture key phytocompounds.

A range of plant defence mechanisms, such as the production of ROS, the oversensitive rejoinder, the production of antimicrobial, phytoalexin secondary chemicals, and other symptoms of distrustful character, can be induced through elicitors (Montesano et al., 2003). The loudening of more than a few advantageous bioactive substances, including artemisinin, azadirachtin, and tanshinones was observed to be triggered through the YE elicitor (Zhao et al., 2010). However, the elicitation method varied amongst dissimilar plant species. In most cases, an "elicitor-receptor" complex developed, and a wide range of biochemical reactions was evident. The embryogenic tissue generated from protoplasts had an elevated amount of alkaloid, and under YE-adjusted circumstances, the production of VC and VB was even higher. The activity of YE has previously been linked to the activation of the tryptophan decarboxylase and strictosidine synthase encoding genes STR and TDC, which are involved in the manufacture of terpenoid indole alkaloids (TIA) (Pauwet al., 2004). In previously investigation, the semi-synthetic combination rejoinder period of the synthesis of monomeric indole alkaloids catharanthine and vindoline to create vinblastine was productively accomplished through enzymatic or chemical means.

The study is significant since *Catharanthus* is the only source of pricey, significant chemotherapeutic medicines with anticancer potential, which are present in very tiny amounts (0.0005% DW). As a result, there is an enrichment of alkaloids in this plant. Rosmaric acid (RA) and phenolic substances were both synthesised in *Salvia miltiorrhiza* hairy roots in response to YE and Ag+, an abiotic elicitor. *Orthosiphon aristatus* and Coleus blumei, two other plants under investigation, also showed elicitor-induced RA accumulation (YE and MJ). When YE in suspension was added to L. erythrorhizon, it quickly and momentarily increased RA accumulation and caused Silybum marianum to produce three times as much silymarin. *Gymnema sylvestre* was shown to respond best to elicitation when YE was added at a concentration of 0.5 g/l. Within 20 days of culture, gymnemic acid production increased by over 5.25 times. In *Astragalus chrysochlorus* cells treated with YE, there was a reported rise in total phenolic content of over (113.3%) compared to control. YE treatments resulted in enriched VC and VB alkaloid acquiesce, and it was found to be 1.5 g/l as the perfect dose (Madrid et al., 2010).

Conclusion

Vinblastine and vincristine are drugs for anticancer that perform through binding to intracellular tubulin in tumour cells, inhibiting the RNA synthesis and DNA repair mechanisms, and DNA-reliant RNA polymerase is blocking. *Catharanthus* is the only source of costly, significant chemotherapeutic medications with anticancer potential.

References

Almagro L, Gutierrez J, Pedreño MA, Sottomayor M. 2014. Synergistic and additive influence of cyclodextrins and methyl jasmonate on the expression of the terpenoid indole alkaloid pathway genes and metabolites in *Catharanthus roseus* cell cultures. *Plant Cell Tissue Organ Cult.*, 119, 543–551.

Almagro L, Sabater-Jara AB, Belchi-Navarro S, Fernandez-Perez F, Bru R, Pedreño MA. 2011. Effect of UV light on secondary metabolite biosynthesis in plant cell cultures elicited with cyclodextrins and methyljasmonate. In *Plants and Environment*, 1st ed.;

Vasanthaiah, H. K. N., Kambiranda, D., Eds.; InTech Janeza Trdine: Rijeka, Croatia, 1, 115–136.

Aslam J, Khan SH, Siddiqui ZH, Fatima Z, Maqsood M, Bhat MA, Nasim SA, Ilah A, Ahmad IZ, Khan SA, Mujib A. 2010. *Catharanthus roseus* (L.) G. Don. An important drug: it's applications and production, *Pharm. Glob.*, 4(12), 1-6.

Aslam J, Mujib A, Nasim SA, Sharma MP. 2002. Screening of vincristine yield in ex *vitro* and *in vitro* somatic embryos derived plantlets of *Catharanthus roseus* L. (G) Don. *Sci. Hortic.*, 119, 325-329.

Bagnyukova TV, Serebriiskii IG, Zhou Y, Hopper-Borge EA, Golemis EA, Astsaturov I. 2010. Chemotherapy and signaling: how can targeted therapies supercharge cytotoxic agents? *Cancer Biol. Ther.*, 10, 839-853.

Barthe L, Ribet JP, Pelissou M, 2002. Optimization of the separation of Vinca alkaloids by nonaqueous capillary electrophoresis. *J. Chromatogr. A*, 968 (1-2), 241-250.

Boller T, Felix G. 2009. A renaissance of elicitors: perception of microbe-associated molecular patterns and danger signals by pattern recognition receptors. *Annu. Rev. Plant Biol.*, 60, 379-406.

Carqueijeiro I, Noronha H, Duarte P, Geros H, Sottomayor M. 2013. Vacuolar transport of the medicinal alkaloids from *Catharanthus roseus* is mediated by a proton-driven antiport. *Plant Physiol.*, 162, 1486–1496.

Das S, Sharangi AB. 2017. Madagascar periwinkle (*Catharanthus roseus* L.): Diverse medicinal and therapeutic benefits to humankind. *J. Pharmacogn. Phytochem.*, 6(5), 1695–701.

Deus-Neumann B, Stöckigt J, Zenk MH. 1987. Radioimmunoassay for the quantitative determination of catharanthine. *Planta Med.*, 53(2), 184-188.

Goldhaber-Pasillas GD, Mustafa NR, Verpoorte R. 2014. Jasmonic Acid Effect on the Fatty Acid and Terpenoid Indole Alkaloid Accumulation in Cell Suspension Cultures of *Catharanthus roseus. Molecules*, 19, 10242–10260.

Gupta MM, Singh DV, Tripathi AK, Pandey R, Verma RK, Singh S, Shasany AK, Khanuja SPS. 2005. Simultaneous Determination of Vincristine, Vinblastine, Catharanthine, and Vindoline in Leaves of *Catharanthus roseus* by High-Performance Liquid Chromatography. *J. Chromatogr. Sci.*, 43(9), 450-453.

Hallard D, van der Heijden R, Verpoorte R, Lopez-Cardoso I, Pasquali G, Memelink J, Hoge JHC. 1997. Suspension cultured transgenic cells of Nicotiana tabacum expressing tryptophan decarboxylase and strictosidine synthase cDNAs from Catharanthus roseus produce strictosidine upon feeding of secologanin. *Plant Cell Rep.*, 17, 50–54.

Horvath P, Ivanyi G. 1982. Quantitative analysis of natural drugs III. Densitometric determination of vinblastine and other alkaloids of *Catharantus roseus. Acta Pharm. Hung.*, 52(4), 150-157.

Jackson DV, Castle MC, Bender RA. 1978. Biliary excretion of vincristine. *Clin. Pharmacol. Ther.* 24, 101–107.

Jaggi M, Kumar S, Sinha AK. 2011. Overexpression of an apoplastic peroxidase gene CrPrx in transgenic hairy root lines of *Catharanthus roseus. Appl. Microbiol. Biotechnol.*, 90, 1005–1016.

Kabesh K, Senthilkumar P, Ragunathan R, Kumar RR. 2015. Phyto-chemical analysis of Catharanthus roseus plant extract and its antimicrobial activity. *Int. J. Pure App. Biosci*., 3(2), 162-72.

Kuboyama, T., Yokoshima, S., Tokuyama, H. and Fukuyama, T. 2004. Stereocontrolled total synthesis of (+)- Vinblastine. *Pure Appl. Chem.,* 75, 29-38.

Kumar A. 2016. Vincristine and vinblastine: a review. *Int. J. Med. Pharm. Sci.*, 6(1), 23-30.

Li M, Peebles CA, Shanks JV, San KY. 2011. Effect of sodium nitroprusside on growth and terpenoid indole alkaloid production in *Catharanthus roseus* hairy root cultures. *Biotechnol. Prog*., 27, 625–630.

Lin Z, Qing-Hui G, Yuan-Gang Z, Lei Y, Yu-Liang M, Yang L. 2014. Simultaneous quantitative determination of five alkaloids in *Catharanthus roseus* by HPLC-ESI-MS/MS. *Chinese J. Nat. Med.*, 12(10), 786-793.

Madrid E, Corchete P. 2010. Silymarin secretion and its elicitation by methyl jasmonate in cell cultures of *Silybum marianum* is mediated by phospholipase D-phosphatidic acid. *J. Exp. Bot*., 61(3), 747-754.

Mu FS, Yang L, Wang W. 2012. Negative-pressure cavitation extraction of four main vinca alkaloids from *Catharanthus roseus* leaves. *Molecules*, 17 (8), 8742-8752.

Naaranlahti T, Auriola S, Lapinjoki SP. 1991. Growth related dimerization of vindoline and catharanthine in *Catharanthus roseus* and effect of wounding on the process. *Phytochemistry*, 30, 1451–1453.

Noble RL. 1990. The discovery of the Vinca alkaloids – chemotherapeutic agents against cancer. *Biochem. Cell Biol*., 68(12), 1344-1351.

Okouneva T, Hill BT, Wilson L, 2003. The effects of vinflunine, vinorelbine, and vinblastine on centromere dynamics. *Mol. Cancer Ther.*, 2(5), 427-436.

Owellen RJ, Hartke CA, Hains FO. 1977. Pharmacokinetics and metabolism of vinblastine in humans. *Cancer Res*., 37, 2597–2602.

Owellen RJ, Hartke CA. 1975. The pharmacokinetics of 4-acetyl tritium vinblastine in two patients. *Cancer R*es., 35, 975–980.

Peebles C, Sander GW, Hughes EH, Peacock R, Shanks JV, San KY. 2011. The expression of 1-deoxy-D-xylulose synthase and geraniol-10-hydroxylase or anthranilate synthase increases terpenoid indole alkaloid accumulation in *Catharanthus roseus* hairy roots. *Metab. Eng*., 13, 234–240.

Pennanen S, Huhtikangas A. 1990. Photochemical one-pot synthesis of vinblastine and vincristine. *Photochem. Photobiol.*, 51(5), 515-518.

Raina SK, Wankhede DP, Jaggi M, Singh P, Jalmi SK, Raghuram B, Sheikh AH, Sinha AK. 2012. CrMPK3, a mitogen activated protein kinase from Catharanthus roseus and its possible role in stress induced biosynthesis of monoterpenoid indole alkaloids. *BMC Plant Biol*., 12, 134.

Rosenthal S, Kaufman S. 1974. Vincristine neurotoxicity. *Ann. Intern. Med*., 80, 733–737.

Ruiz-May E, Galaz-Ávalos RM, Loyola-Vargas VM. 2009. Differential secretion and accumulation of terpene indole alkaloids in hairy roots of *Catharanthus roseus* treated with methyl jasmonate. *Mol. Biotechnol*., 41, 278–285.

Sahi N, Mostajeran A, Ghanadian M. 2022. Changing in the production of anticancer drugs (vinblastine and vincristine) in *Catharanthus roseus* (L.) G. Don by potassium and ascorbic acid treatments. *Plant Soil Environ.,* 68, 18–28.

Sertel S, Fu YJ, Zu YG, 2011. Molecular docking and pharmacogenomics of Vinca alkaloids and their monomeric precursors, vindoline and catharanthine. *Biochem Pharmacol*, 81 (6): 723-735.

Sethi VS, Kimball JC. 1981. Pharmacokinetics of vincristine sulfate in children. *Cancer Chemother. Pharmacol.* 6, 111–115.

Sottomayor M, Barceló AR. 2006. The Vinca alkaloids: From biosynthe-sis and accumulation in plant cells, to uptake, activity and metabolism in animal cells. In: *Studies in Natural Products Chemistry*. Elsevier, pp. 813–857.

Tam A, Gotoh H, Robertson WH, Boger DL. 2010. Catharanthine C16 substituent effects on the biomimetic coupling with vindoline: Preparation and evaluation of a key series of vinblastine analogues. *Bioorg. Med. Chem. Lett*., 20, 6408-6410.

Tikhomiroff C, Jolicoeur M. 2002. Screening of *Catharanthus roseus* secondary metabolites by high-performance liquid chromatography. *J. Chromatogr. A*, 955(1), 87-93.

Van den Berg HW, Desai ZR, Wilson R, Kennedy G, Bridges JM, Shanks RG. 1982. The pharmacokinetics of vincristine in man: reduced drug clearance associated with raised serum alkaline phosphatase and dose-limited elimination. *Cancer Chemother. Pharmacol*., 8, 215–219.

Van Der Heijden R, Jacobs DI, Snoeijer W, Hallard D, Verpoorte R. 2004. The Catharanthus alkaloids: pharmacognosy and biotechnology. *Curr. Med. Chem*., 11(5), 607–628.

Yu F, de Luca V. 2013. ATP-binding cassette transporter controls leaf surface secretion of anticancer drug components in *Catharanthus roseus. Proc. Natl. Acad. Sci*., 110, 15830–15835.

Zandi M. 2021. Cytotoxicity of taxol in combination with vincristine and vinblastine against A375 cell line. *Gene Cell Tissue,* 8(4), e114359.

Zhao J, Hu Q, Guo YQ, Zhu WH. 2001. Effects of stress factors, bioregulators, and synthetic precursors on indole alkaloid production in compact callus clusters cultures of *Catharanthus roseus*. *Appl. Microbiol. Biotechnol*., 55, 693–698.

Zhao J, Zhu WH, Hu Q, He XW. 2001. Enhanced indole alkaloid production in suspension compact callus clusters of *Catharanthus roseus*: Impacts of plant growth regulators and sucrose. *Plant Growth Reg*., 33, 33–41.

Zhao J, Zhu WH, Hu Q. 2001. Enhanced catharanthine production in Catharanthus roseus cell cultures by combined elicitor treatment in shake flasks and bioreactors. *Enzym. Microb. Technol*., 28, 673–681.

Zhou XJ, Rahmani R. 1992. Preclinical and clinical pharmacology of Vinca alkaloids. *Drugs*, 44(Suppl 4), 66-69.

Zhou XJ, Zhou-Pan XR, Gauthier T, Placidi M, Maurel P, Rahmani R. 1993. Human liver microsomal cytochrome P450 3A isozymes mediated vindesine biotransformation. Metabolic drug interactions. *Biochem. Pharmacol*., 45, 853–861.

Zhou-Pan XR, Seree E, Zhou XJ. 1993. Involvement of human liver cytochrome P450 3A in vinblastine metabolism: drug interactions. *Cancer Res*., 53, 5121–5126.

Chapter 6

Stevensine: A Bromopyrrole Alkaloid from Marine Sponges

Ankit Kumar[1,2,*], Rekha Jethi[3], Ravindra Semwal[1], Ganesh Kumar[2] and Deepak Kumar Semwal[4]

[1]Research and Development Centre, Faculty of Biomedical Sciences, Uttarakhand Ayurved University, Dehradun, India
[2]College of Pharmacy, Shivalik campus, Dehradun, India
[3]Patanjali Research Institute, Patanjali Yogpeeth, Haridwar, Uttarakhand, India
[4]Department of Phytochemistry, Faculty of Biomedical Science, Uttarakhand Ayurved University, Dehradun, India

Abstract

The dependency of humankind on nature for food and medicine, since time immemorial, has continued so far. Plants, animals, micro-organisms and marine organisms are excellent sources of bioactive molecules like alkaloids, glycosides, steroids, tannins, flavonoids, terpenoids and saponins etc. Due to the efficacy and potency of the naturally derived molecules, at present, no alternative is available for natural molecules, such as morphine, digitoxin, quinine, penicillin and cephalosporin. Due to the development of resistance against current medication in the case of micro-organism infection, parasite infection and cancer, the need for a new compensatory generation of therapeutic agents is today's demand. Other than plants, animals and microorganisms, marine life has been introduced as an alternative source of medicinal agents in the last 50 years. At present, more than 200,000 marine species, as representative of

* Corresponding Author's Email: ankitkumarkoli88@gmail.com.

In: The Essential Guide to Alkaloids
Editor: Deepak Kumar Semwal
ISBN: 979-8-88697-456-0

each *Phylum*, are found in the ocean. About 30,000 bioactive compounds have been discovered in marine organisms. More than 8,500 species of sponges, are scientifically documented and identified but Researchers believed more than 25,000 species exist in the ocean. As a source of medicine, about 5,300 bioactive compounds are reported in marine sponges from diverse chemical classes, like alkaloids, peptides, and terpenoids. This book chapter discussed biological source, synthesis and pharmacological studies of a bromopyrrole alkaloid "Stevensine" isolated from different species of marine sponge. It was also synthesized from the related marine alkaloid hymenin. Stevensine has been shown several pharmacological potentials in different *in-vitro* and *in-vivo* studies, such as protein kinase inhibitor, neurological, anti-microbial, anti-tubercular, anti-cancer and anti-parasite activity. The previously available report concludes that further studies are needed to explore the mechanism of action of stevensine in cancer, diabetes and neurological disorders.

Keywords: alkaloids, marine compounds, sponges, protein kinase inhibitory

1. Introduction

Since time immemorial, as a primary source of medicine, humankind has depended on nature and has continued so far. Natural sources cover living and non-living things both provide bioactive compounds in various forms. As a living natural source (animal, plant, fungi, protist and monera), medicinal substances are obtained from organised (whole or their parts like cells, tissue and organ) and unorganised source (their extract, exudate and produced biochemical). More than 50% of marketed medicinal substances are either obtained from natural sources or formed by the synthetic process using natural substances as a starting material (Fenical, 1999).

For the treatment of a large variety of diseases via novel mechanisms, the need for new therapeutic agents is a continuous inspiration for drug discovery. Through the available bioactive compounds from different sources such as plants, animals and microorganisms, the successful treatment of various forms of cancer, fungal and viral infections, inflammatory diseases, and neurodegenerative diseases is not possible yet. The development of resistance against existing medication is a major issue in the case of micro-organisms/parasite infection and cancer, which requires the compensatory generation of new drugs. In the past few years, marine life has been introduced as the latest source of therapeutically active compounds.

The planet is divided into five separate oceans (Pacific, Atlantic, Indian, Arctic, and Southern/Antarctic) that cover about 71% of Earth's surface, but these oceans all connect into a single world ocean (Farmer and Cook, 2013). At present, more than 200,000 marine species, as a representative of each *Phylum*, are found in the ocean and have been reported that range from microscopic levels (like phytoplankton) to huge cetaceans (like the blue whale) including protists, bacteria and their associated viruses. Marine micro-organisms have been estimated as constituting about 70%of the total marine biomass (Bar-On et al., 2018). About 30,000 bioactive compounds are discovered in marine organisms (Kiuru et al., 2014). In the past 50 years, natural products from marine sources have been explored as a field of research to develop several novel therapeutic agents (Faulkner, 1991). Due to our curiosity about the natural world, SCUBA and submersible techniques were developed to provide the facility for human beings to penetrate the earth's aquatic environment (Scheuer, 1991). During this period, various marine lives have been discovered which are reported to provide nutrient-rich food material as well as medicinal substances. Further research has proved the importance of marine organisms as a source of novel bioactive molecules. Secondary metabolites obtained from marine sources provide a wealth of compounds. About 8,500 species of Sponges are scientifically documented and identified but scientists/researchers believed more than 25,000 species exist in the ocean. Sponges are the world's simplest multicellular organisms that are found in different shapes, sizes, colours and textures but do not have digestive, nervous, or circulatory systems. They have some characteristic features like a network of pores on their body plan, choanocyte chambers and channels that allow water to circulate through them (Rinehart, 2007). They are classified in the Phylum *Porifera*, which has four distinct classes- *Demospongiae* (containing 90% of all living sponges), *Calcarea* (calcareous sponges), *Hexactinellida* (rare glass sponges) and *Homoscleromorpha* (Renard et al., 2013). Due to the amazing diversity of therapeutically bioactive compounds, marine sponge attracts the extensive attention of pharmaceutical professionals, like Chemists and Pharmacologists. More than 5,300 bioactive compounds are reported in diverse chemical classes, like alkaloids, peptides, and terpenoids from different marine sponges, such as halichondrins, discodermolide, hemiasterlins, and arenastatin (Ebada and Proksch, 2012). This book chapter illustrates the biological source, biosynthesis/synthesis, and important pharmacological studies of a bioactive alkaloid Stevensine.

2. Biological Source and Isolation of Stevensine

Albizatit and Faulkner (1985) isolated a novel marine alkaloid "Stevensine" from the Micronesian sponge (76.1 g dry weight) that was collected from (Scuba, -15 m) at Ponape, Carolina Islands by hand. The collected specimen was soaked for six months at 0°C in methanol. After six months, the amorphous light orange solid was obtained as a residue by solvent evaporation and the residue was further sequentially treated with hexane, Dichloromethane, Ethyl acetate and methanol. The methanol fraction was chromatographed on a crosslinked dextran-based resin Sephadex LH-20 with methanol as eluent. The collected fraction was again rechromatographed by using silica (stationary phase) and Methanol: dichloromethane and concentrated Ammonium hydroxide (6:3:1) as a solvent system. From the active fraction (antimicrobial assay), stevensine (0.10% dry weight) was obtained. Chemically, it has the 6,7-dihydropyrrolo[2,3-c]azepin-8-one ring system. It is an amorphous solid, orange in colour and soluble in polar solvents. On heating, it decomposed slowly. Stevensine is a bromopyrrole alkaloid (Mohammed et al., 2006) (Figure 1), also known as Odiline (Zhang, et al., 2012). Stevensine has the molecular formula $C_{11}H_9N_5OBr_2$ (Mohammed et al., 2006). The alkaloid "Stevensine" is named for the late Robert V. Stevens (Wayne State University, Detroit, Michigan) (Albizati and Faulkner, 1985). In another study, Stevensine was obtained from the Jamaican sponge *Stylissa caribica* (deep orange in colour) which was collected from a silted reef (22-30 m deep) at Columbus Park, Jamaica. The fresh sponge was dissolved in ethanol and the extract was subjected to silica gel vacuum-liquid chromatography followed by column chromatography, Sephadex LH-20, Reversed-Phase High-Performance Liquid Chromatography, and Preparative thin layer chromatography to yield Stevensine as a yellowish powder (Mohammed et al., 2006). Furthermore, it was reported from a number of sponges e.g., from the *Axinella* species (*A. corrugate* and *A. verrucose*), *Pseudaxinyssa* species, *Stylissa* species (*S. caribica* and *S. flabelliformis*) and some unidentified species (Albizatit and Faulkner, 1985; Newbold et al., 1999; Aielloet al., 2006; Nguyen and Tepe, 2009; Fouad et al., 2012; Zhang et al., 2012; Ebada et al., 2014; Vitale et al., 2018; Hafez Ghoran and Kijjoa, 2021).

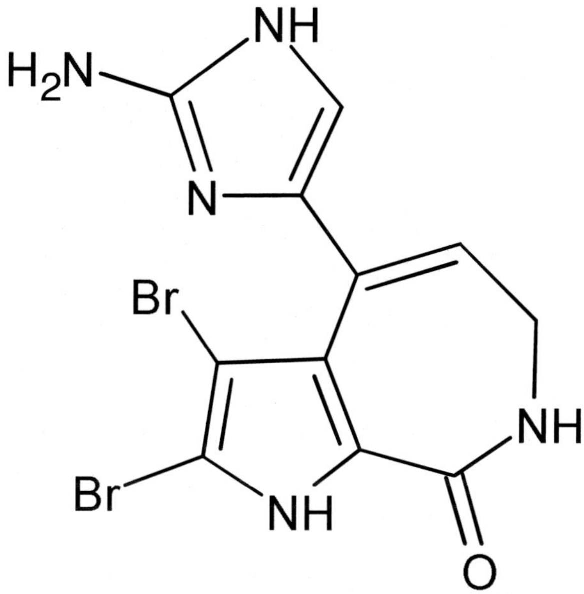

Figure 1. Chemical structure of Stevensine (IUPAC name: 4-(2-Amino-1*H*-imidazol-5-yl)-2,3-dibromo-6,7-dihydropyrrolo[2,3-c]azepin-8(1*H*)-one).

3. Synthesis of Stevensine

Xu et al. (1996) reported the synthesis of stevensine from the related alkaloid, hymenin. As an alkaloid, Hymenin was isolated from an *Hymeniacidon* species (Okinawan sponge) in 1984 (Kobayashi et al., 1986). In this study, the direct bromination of Hymenin in trifluoroacetic acid (CF_3CO_2H) gives 4'-bromohymenin. Upon heating at 90°C for 24 hr in methanesulfonic acid (CH_3SO_3H) in an unsealed flask, 4'-bromohymenin produced both hymenin (40%) and stevensine (20%) (Figure 2). The author suggested that the disproportionation-like mechanism involves in the production of hymenin and stevensine. Xu et al. (1997) reported another synthesis of stevensine. Andrade et al. (1999) described the biosynthesis of stevensine, in a cell culture composed of archaeocytes of the marine sponge *Teichaxinella morchella*.

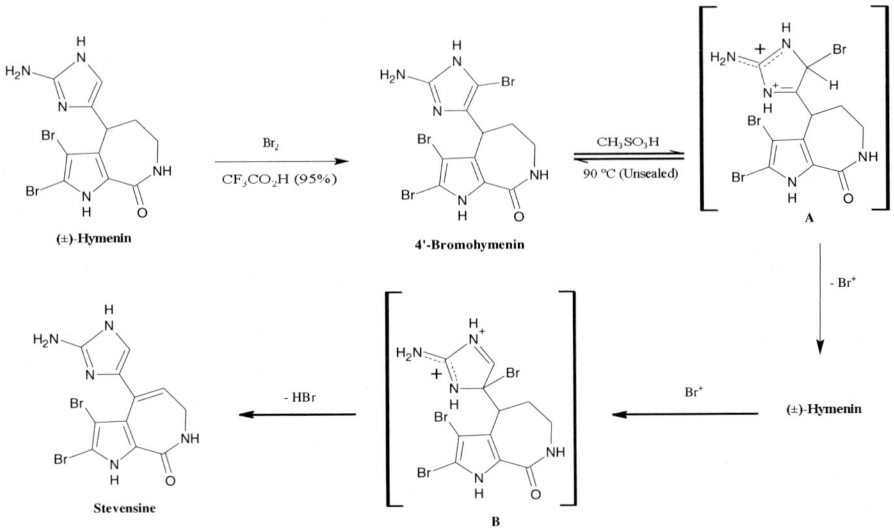

Figure 2. Chemical synthesis of Stevensine.

4. Pharmacological Activities of Stevensine

4.1. Protein Kinase Inhibitory Activity

The agents that regulate the action of enzymes are known as protein kinases. Several types of protein kinases are well defined that control and regulate the different types of cell functions such as cell signalling, cell growth and cell division. More than five hundred protein kinase genes are encoded by the human genome. Research revealed that secondary metabolites which have an action on protein kinase activity are developed as therapeutic molecules to treat several pathological disorders (Anshabo et al., 2021). Stevensine, isolated from methanolic extract of the marine sponge *Stylissa flabelliformis* (collected from South Sulawesi, Indonesia) was tested as an inhibitor of a different type of kinase that plays an important role in the treatment of cancer, diabetes, neurodegenerative disorders etc. This study included the kinase i.e., glycogen synthase kinase-3 (GSK-3), CDC-like kinase 1 (CLK-1), cyclin-dependent kinase 5 (CDK5), dual-specificity tyrosine-(Y)-phosphorylation-regulated kinase 1A (DYRK1A), casein kinase 1 (CK-1), Cyclin-dependent kinase 1 (CDK1), cyclin-dependent kinase 2 (CDK2/A), complex of cyclin-dependent kinase 9 and cyclin T (CDK9/cyclin T) and *Plasmodium falciparum* glycogen synthase kinase-3 (PfGSK-3) (Ebada et al., 2014). Pharmacologically, the inhibitors of glycogen synthase kinase-3 (GSK-3) are an important strategy in the treatment of neurodegenerative and psychiatric disorders. The cation lithium was discovered as the first natural GSK-3 inhibitor and proved to be a neuroprotective agent (Eldar-Finkelman and Martinez, 2011). The therapeutic potential of CDC-like kinases (CLKs) has been explored mainly over the last decade. CLKs are an attractive target that covers a wide range of therapeutic areas. Studies evidenced that the inhibitors of CLK play an important role as an effective agent in the treatment of cancer, frontotemporal dementia, Parkinson's disease, tauopathies, diabetes, Phelan-McDermid syndrome and malaria etc. (Martín Moyano et al., 2020). During the process of cell division, it is controlled by a complex composed of CDKs. The main involvement of CDKs is in the cell cycle but it is also responsible for other important cellular processes, such as transcription of genes, insulin secretion, neuronal functions and glycogen synthesis etc. (Zhang et al., 2021). CDK1 inhibitors are developed as an anti-cancer agent that induces prolonged G2 arrest and sensitizes cells to DNA damaging agents in cancer cells, resulting in cell death (Sunada et al., 2021). The selective inhibitors of CDK2 may also provide a therapeutic benefit against certain cancers and may be developed as an anti-cancer drug (Tadesse et al., 2018). During mammalian embryogenesis, CDK5 plays an important role in the development of the central nervous system (CNS), but in adults, it is essential for the proper functioning of neuronal architecture. The deregulation of CDK5 causes cytotoxic effects and produces neurodegenerative disorders (Mapelli and Musacchio, 2003). The overexpression of CDK9 causes cancer in humans, especially in Triple-negative breast cancer. Studies proved the anti-proliferative and anti-migration role of CDK9-cyclin T1 protein inhibitors in breast cancer cell lines (Cheng et al., 2022). DYRK1A has diverse effects and functions on cell regulation. Due to its abnormal expression and activity, several abnormalities such as malignancies, Down syndrome, and Alzheimer's disease are reported. Studies reported the effectiveness of DYRK1A as an important therapeutic agent for diabetes due to having a key role in pancreatic β-cell proliferation (Yoon et al., 2020). CK-1 phosphorylates the key regulatory proteins that are involved in the control of cell functioning and response, such as differentiation, proliferation, segregation of chromosomes, DNA repair, circadian rhythms, protein translation, immune response, inflammation and metabolic pathways etc. The inhibitors of CK-1 act by various mechanisms and play an important role in therapeutic action for various poorly practicable clinical conditions (Janovská et al., 2020). The eukaryotic protein kinase PfGSK-3 inhibitor has been suggested as the reputed target for novel antimalarial drugs. However, its physiological functions are still unknown (Masch and Kunick, 2015). The study conducted by Ebada et al. (2014) showed the effectiveness of stevensine against 9 different protein kinases. It was found active against all tested protein kinases except PfGSK-3. The IC_{50} was recorded as 0.39, 0.36, 0.49, 2.5, 0.78, 0.31, 2.2 and 0.9 μM for GSK-3, CLK-1, CDK1, CDK2/A, CDK5, CDK9/cyclin T, DYRK1A and CK-1, respectively (Ebada et al., 2014). Stevensine, isolated from *Axinella* sp. of sponges was tested as a therapeutic agent of neurodegenerative disease via targeting β- secretase, casein kinase 1, cyclin-dependent kinase 5 and glycogen synthase kinase 3β. It was found inactive as kinase inhibitory up to 10μM concentration (Zhang et al., 2012).

4.2. Neurological Activity

Globally, about 12% of total deaths are caused by neurological disorders that cause severe health challenges in developing countries, which are caused by dysfunction or abnormalities in the nervous system that result in psychological or physical symptoms. Alzheimer's disease, cerebrovascular migraine disease, Japanese encephalitis, Parkinson's disease, tetanus, dementias, neuropathies, meningitis, poliomyelitis and multiple sclerosis are common neurological disorders. Alzheimer's and other dementias caused about 0.34, 0.41, 0.46 and 2.84% of the total deaths in the lower middle, low, upper middle, and high-income countries, respectively, in 2005 (Semwal et al., 2021). A sponge, *Axinella verrucose,* was collected from the Gulf of Naples to obtain Stevensine. In a study based on computational approaches i.e., molecular docking and molecular dynamics (MD) simulations, it was recorded as an inhibitor of proteins responsible for Alzheimer's. Stevensine was found as a possible inhibitor of acetylcholinesterase (AChE), butyrylcholinesterase (BuChE), and β-site amyloid precursor protein-cleaving enzyme 1 (BACE-1). The inhibitors of these proteins are targeted as anti- Alzheimer's drugs. The Stevensine binds to the catalytic anionic site (CAS) of human AChE. Moreover, modelling shows a loosely binding mode for CAS of equine BuChE. It was able to inhibit and restore β-amyloid (Aβ) aggregation of both Aβ1–40 and Aβ1–42 peptides. The biological activity of Stevensine was confirmed by performing biochemical assays. Stevensine exhibited higher activity than other tested compounds in the study. Stevensine inhibits the murine BACE-1 with IC_{50} values of 1.4 μM (Hafez Ghoran and Kijjoa, 2021). Stevensine, isolated from the Caribbean sponges *Stylissa caribica* and *Agelas wiedenmayeri* was evaluated for interactions with voltage-induced cellular calcium homeostasis using PC12 cells. Stevensine showed no detectable effect when applied at a concentration of 10 μM. If the metabolite has the efficacy to disturb cellular calcium signalling in neuroendocrine PC12 cells, it shows its neurotoxic potential (Bickmeyer et al., 2007).

4.3. Antitubercular Activity

Several synthetic molecules have been reported for their effect against tuberculosis, caused by the infection of *Mycobacterium tuberculosis*. It is an acid-fast aerobic bacterium that can grow on gram stain as either gram-positive or gram-negative. The United States Food and Drug Administration approved the combination therapy of synthetic medication as a first-line drug (isoniazid, rifampin, pyrazinamide and ethambutol), second-line drug (streptomycin, capreomycin, amikacin, levofloxacin, moxifloxacin and gatifloxacin) and multi-drug-resistant tubercular drug for the treatment of tuberculosis. Due to the adverse effect of available synthetic medication, naturally derived molecules are always an attractive area of research in pharmaceutical science (Umumararungu et al., 2020). Like bengamides P 2, bengamides Q 3 (Acquah et al., 2022), fistularin-3, 11- deoxyfistularin-3 (de Oliveira et al. 2006), ileabethoxazole (Rodríguez et al. 2006) and manzamine alkaloid (Rao et al., 2006) are several examples of anti-tubercular compound isolated recently from the Marine organism. Stevensine, isolated from the Jamaican sponge *Stylissa caribica* was tested as an inhibitor of *Mycobacterium tuberculosis*. Stevensine showed 95% inhibition against tested micro-organisms at 128 μg/mL (Mohammed et al., 2006).

4.4. Anti-Cancer and Cytotoxic Activities

Anti-cancer and cytotoxic potentials of about 136 and 94 novel natural compounds, respectively, have been reported from Marine organisms (invertebrate animals, fungi, bacteria and algae that belong to diverse chemical classes such as polyketides, peptides, terpenes and steroids). Most of the compounds have defined pharmacological action in a number of experimental and clinical models with mechanisms of action. Examples of these compounds are ara-A-C, agosterol, amphidinolides, ankaraholides, aplyronine, aptamine, ascididemin, bastadin, bastadin, blumiolides, bromovulone, clavulone, cryptophycins, dideoxypetrosynol, dolastatins, fascaplysin, gymnastatins, hemiasterlin, homaxisterol, kahalalide, lamellarin, leiodolides, marinomycins, melophlin, mycapolyols, philinopsides, psammaplin, seragamides, stellettins, variolin, verrucarin, and ygosporamid (Mayer and Gustafson, 2008). Stevensine, isolated from Jamaican sponge *Stylissa caribica* has shown no cytotoxicity against a number of cancer cell lines (Mohammed et al., 2006). The Indonesian marine sponge *Stylissa* species was collected from the Derawan Islands, Indonesia in 2008 by scuba method. Stevensine was obtained from the methanol extracts by chromatography and fractionation. The isolated Stevensine was tested for cytotoxic effect against L5178Y mouse lymphoma cells of the rats at the concentration of 10 µg/mL by using a micro-culture tetrazolium (MTT) assay. Kahalalide F was used as a control and showed IC_{50} at 6.3 µg/mL while the Stevensine showed no activity (< 10% inhibition) up to the tested concentration (Fouad et al., 2012). The cytotoxicity of Stevensine was against a primary cell line that derived from rat skeletal myoblasts named L6-cells tested using a 96-well microtiter plate method. The CC_{50} value of Stevensine was determined as >90 µg/ml and 0.005 µg/ml for podophyllotoxin, which was used as the standard drug (Scala et al., 2010).

4.5. Anti-Microbial Activity

In 1928, Alexander Fleming discovered the first antibiotic (penicillin) at St. Mary's Hospital, London, since thousands of structurally diverse antimicrobial compounds are reported from a variety of natural organisms. The anti-microbial compound plays a role as a lifesaving drug that fights against various types of infections caused by pathogenic microorganisms every year (Buynak, 2004). Cephalosporins, a beta-lactam antimicrobials compound for the first time in 1945 from the marine fungus *Cephalosporium acremonium* that is used in the treatment of a wide range of infections caused by gram-negative and gram-positive bacteria (Fiore et al., 2022). Instamycin, an aminoglycoside isolated from the marine micro-organism *Streptomyces tenjimariensis,* has been reported as an antibiotic (Shah and Seth, 2014). More than 300 marine actinomycetes-derived compounds are reported up to 2019 as anti-microbial agents from diverse chemical classes such as polyketides, terpenoids, nitrogen-containing compounds and sterols etc. (Wang et al., 2020). Stevensine isolated from Caribbean reef sponge *Axinella corrugata* that was collected from the Bahamas and the Florida Keys, USA against a test panel of marine bacteria *Vibrio parahaemolyticus* (ATCC 27969), Vibrio parahaemolyticus (isolated from healthy mangrove sponge *Chondrilla nucula*), *Vibrio alginolyticus* (isolated from healthy reef sponge *Pandaros acanthifolium*), *Deleya marina* (ATCC 25374), two Bacillus species (isolated from the surface tissues of necrotic reef sponges- *Ircinia strobilina* and *Agelas clathrode*), *Listonella anguillarum* (from seawater) and *Alcaligenes piechaudii* (sea water) by using disc-diffusion assay at the concentrations of 10-200 mg/ml. Gentamicin is used as a standard anti-biotic. At the concentration of 50-200 mg/ml, the stevensine was found active against *V. alginolyticus*, *Deleya marina* and both species of *Bacillus*. However, the extract of sponge showed greater activity than Stevensine but at 200 mg/ml Stevensine inhibition was recorded higher against *Bacillus* species (isolated from the surface tissues of necrotic reef sponges *Agelas clathrode)* (Newbold et al., 1999). Stevensine, isolated from the Jamaican sponge *Stylissa caribica,* was tested for antimicrobial effects against yeast (*Candida albicans*, *Cryptococcus neoformans* and *Aspergillus fumigatus*), Gram-positive bacteria (*Staphylococcus aureus*, Methicillin-resistant *Staphylococcus aureus*, *Mycobacterium intracellulare*) and Gram-negative bacteria (*Pseudomonas aeruginosa*). The Stevensine was found inactive against all tested microorganisms (Mohammed et al., 2006). Growth inhibitory activity of Stevensine, isolated from *Axinella* sp. of sponges, was screened against Gram positive (*Staphylococcus aureus* ATCC 9144 and 25923 and *Bacillus subtilis* ATCC 6051 and 6633) bacteria, Gram negative (*Escherichia coli* ATCC 11775 and *Pseudomonas aeruginosa* ATCC 10145) bacteria and the fungus (*Candida albican* ATCC 90028). Stevensine was found inactive against all tested microorganisms (Zhang et al., 2012).

4.6. Antimalarial Activity

In 2020, about 241 million cases of malaria were estimated worldwide. The reported number of total deaths caused by malaria was 627000 in 2020 (World Health Organization, 2022). At present, no single drug has been available that can eradicate all forms of malaria's life cycle (plasmodium parasite). Four major classes of anti-malarial drugs are currently used to treat malarial infection include aryl amino alcohols (quinine, quinidine, chloroquine, amodiaquine), antifolates (pyrimethamine, proguanil, chlorproguanil, trimethoprim), artemisinin derivatives (artemisinin, dihydroartemisinin, artemether, artesunate) and antimicrobials (Arrow et al., 2004). Fattorusso and Taglialatela-Scafati (2009) reported more than 60 compounds from diverse chemical classes obtained from the marine source for their anti-malarial potentials, such as alkaloids (manzamine, cyclostellettamines and halicyclamines), isonitrile-containing derivatives (isonitrile, isothiocyanate, axamide and axisonitrile), polyketide (plakortin, dihydroplakortin, 3-epiplakortin and plakortide Q), terpenoids (igmosceptrellin), peptides (enturamide, dragomabin, dragonamide and gallinamide) quinones and phenols (xestoquinone and ilimaquinone). Stevensine, isolated from the Jamaican sponge *Stylissa caribica* was tested for antimalarial/anti-parasitic activity against (*Plasmodium falciparum*, D6 and W2 clone). Stevensine showed marginal activity, with IC_{50} values of 1800 ng/mL against the D6 clone (Mohammed et al., 2006). Stevensine, obtained from *Axinella verrucose* tested against erythrocytic stages of *Plasmodium falciparum* was determined by *in vitro* assay using the chloroquine and pyrimethamine-resistant K1 Strain. Chloroquine was used as a standard drug. Stevensine showed an IC_{50} value of 4.88 μg/ml while the standard drug Chloroquine has an IC_{50} value of 0.065 μg/ml. In a further study, Stevensine was tested against three different enzymes (PfFabI, PfFabG, PfFabZ) that were involved in the de novo fatty acid biosynthesis pathway of *Plasmodium falciparum* (PfFAS-II). Stevesine was found inactive in this assay (Scala et al., 2010).

4.7. Anti-Protozoal Activity

As a significant threat to human health, the infection of Protozoan parasites causes more than a million deaths annually. Parasitic protozoans of the genus *Trypanosoma* such as *Trypanosoma brucei gambiense* and *T. brucei rhodesiense* cause African trypanosomiasis (sleeping sickness) which attacks the central nervous system (CNS), leading to behavioural changes, coma, and ultimately, if untreated, death. In 2010, African trypanosomiasis was responsible for around 9000 deaths. At present, no vaccine is available for trypanosomiases and the treatment relies only on chemotherapy. The treatment is selected on the basis of species and stage of infection. Several medications are available to treat the parasitic infection but the available synthetic therapeutic agent has a number of adverse effects and resistance is developed by the parasite against them. Thus, the discovery of novel agents from a natural source is the demand of the present time (Andrews et al., 2014). Stevensine, obtained from *Axinella verrucose* tested for anti-parasitic/anti-protozoal activity against parasitic protozoa *Trypanosoma brucei rhodesiense* (STIB 900 strain), *Trypanosoma cruzi* (Tulahuen strain C2C4) and *Leishmania donovani* (strain MHOM/ET/67/L82) using *in vitro* method. The IC_{50} of Stevensine was determined as 25.34, >90 and 75.86 μg/ml, respectively, for the tested parasite while IC_{50} values for standard drugs melarsoprol, benznidazole and miltefosine were determined as 0.004, 0.312, and 0.206 μg/ml against *T. brucei rhodesiense, T. cruzi* and *L. donovani*, respectively (Scala et al., 2010).

4.8. Anti-Inflammatory Activity

As an anti-inflammatory agent, several compounds from Marine organisms are obtained from different chemical classes such as alkaloids (neoechinulin, chaetoglobosin fex), sesquiterpenoids (flavalin, lochmolins, rumphellol, lemnalol), diterpenes (excavatolide, lobocrasol, crassumol), polysaccharides (sulfated polysaccharides), proteins (lectin), steroids (solomonsterol, sclerosteroid), and other chemical compounds (xanthone derivatives), isolated from marine organisms are found to exhibit anti-inflammatory activity (Cheung et al., 2015). Stevensine, isolated from the Jamaican sponge *Stylissa caribica* was tested for anti-inflammatory activity in rat neonatal microglia and found inactive (Mohammed et al., 2006).

4.9. Anti-HIV Activity

The human immunodeficiency virus (HIV) causes acquired immunodeficiency syndrome (AIDS) which targets the immune system and weakens people's defence against many infections. Globally, 38.4 million people were living with HIV at the end of 2021 according to the WHO (2022) report. Worldwide, about 650000 people have died from HIV-related illnesses in 2021. To date, several classes of compounds such as tannins (diphlorethohydroxycarmalol), polysaccharides (fucans, galactofucan, glucoronogalactan, sulfated galactans), lectins (griffithsin, high-mannose-binding lectin), and diterpenes ((6R)-6-hydroxydichotoma-3,14-diene1,17-dial (DT1) and (6R)-6-acetoxydichotoma-3,14-diene-1,17-dial) have been identified, isolated and reported to have significant anti-HIV potential (Kim and Karadeniz, 2011). Stevensine, isolated from Jamaican sponge *Stylissa caribica* tested for anti-human immunodeficiency virus-1 (HIV-1) in peripheral blood mononuclear (PBM) cells. Stevensine was found inactive against HIV-1 (Mohammed et al., 2006).

Conclusion

Stevensine showed several pharmacological potentials in different *in-vitro* and *in-vivo* studies. It was found as an inhibitor of protein kinases such as GSK-3, CLK-1, CDK1, CDK2/A, CDK5, CDK9/cyclin T, DYRK1A and CK-1 that play important role in the treatment of diabetes, cancer and neurological disorders. Other than kinase inhibitors, several other pharmacological potentials of Stevensine have been reported. In future, the study of Stevensine on different experimental models to explore the area of treatment and to understand the mechanism are needed.

References

Acquah, K. S., Beukes, D. R., Seldon, R., Jordaan, A., Sunassee, S. N., Warner, D. F., & Gammon, D. W. (2022). Identification of antimycobacterial natural products from a library of marine invertebrate extracts. *Medicines,* 9(2), 9.

Aiello, A., D'Esposito, M., Fattorusso, E., Menna, M., Müller, W. E. G., Perović-Ottstadt, S., & Schröder, H. C. (2006). Novel bioactive bromopyrrole alkaloids from the Mediterranean sponge Axinella verrucosa. *Bioorganic & Medicinal Chemistry,* 14(1), 17–24.

Albizati, K. F., & Faulkner, D. J. (1985). Stevensine, a novel alkaloid of an unidentified marine sponge. *Journal of Organic Chemistry,* 50(21), 4163–4164.

Andrade, P., Willoughby, R., Pomponi, S. A., & Kerr, R. G. (1999). Biosynthetic studies of the alkaloid, stevensine, in a cell culture of the marine sponge Teichaxinella morchella. *Tetrahedron Letters,* 40(26), 4775–4778.

Andrews, K. T., Fisher, G., & Skinner-Adams, T. S. (2014). Drug repurposing and human parasitic protozoan diseases. *International Journal for Parasitology: Drugs and Drug Resistance,* 4(2), 95-111.

Anshabo, A. T., Milne, R., Wang, S., & Albrecht, H. (2021). CDK9: a comprehensive review of its biology, and its role as a potential target for anti-cancer agents. *Frontiers in Oncology,* 11, 678559.

Arrow, K. J., Panosian, C., & Gelband, H. (2004). Antimalarial drugs and drug resistance. In : *Saving Lives, Buying Time: Economics of Malaria Drugs in an Age of Resistance.* National Academies Press (US).

Bar-On, Y. M., Phillips, R., & Milo, R. (2018). The biomass distribution on Earth. *Proceedings of the National Academy of Sciences*, 115(25), 6506-6511

Bickmeyer, U., Grube, A., Klings, K.-W., & Köck, M. (2007). Disturbance of voltage-induced cellular calcium entry by marine dimeric and tetrameric pyrrole–imidazole alkaloids. *Toxicon, 50(4), 490–497.*

Buynak, J. D. (2004). The discovery and development of modified penicillin-and cephalosporin-derived β-lactamase inhibitors. *Current Medicinal Chemistry,* 11(14), 1951-1964.

Cheng, S., Yang, G. J., Wang, W., Ma, D. L., & Leung, C. H. (2022). Discovery of a tetrahydroisoquinoline-based CDK9-cyclin T1 protein–protein interaction inhibitor as an anti-proliferative and anti-migration agent against triple-negative breast cancer cells. G*enes & Diseases,* 9(6), 1674-1688.

Cheung, R. C. F., Ng, T. B., Wong, J. H., Chen, Y., & Chan, W. Y. (2015). Marine natural products with anti-inflammatory activity. *Applied Microbiology and Biotechnology,* 100(4), 1645–1666.

de Oliveira, M. F., de Oliveira, J. H., Galetti, F. C., de Souza, A. O., Silva, C. L., Hajdu, E., Peixinho, S., & Berlinck, R. G. (2006). Antimycobacterial brominated metabolites from two species of marine sponges. *Planta Medica,* 72(05), 437-441.

Ebada, S. S., & Proksch, P. (2012). The chemistry of marine sponges∗. *Handbook of marine natural products,* 191-293.

Ebada, S. S., Linh, M. H., Longeon, A., de Voogd, N. J., Durieu, E., Meijer, L., Bourguet-Kondracki, M. L., Singab, A. N., Müller, W. E., & Proksch, P. (2014). Dispacamide E and other bioactive bromopyrrole alkaloids from two Indonesian marine sponges of the genus Stylissa. *Natural Product Research,* 29(3), 231–238.

Eldar-Finkelman, H., & Martinez, A. (2011). GSK-3 inhibitors: preclinical and clinical focus on CNS. *Frontiers in molecular neuroscience*, *4*, 32.

Farmer, G. T., & Cook, J. (2013). The World Ocean. In : *Climate Change Science: A Modern Synthesis* (pp. 247-259). Springer, Dordrecht.

Fattorusso, E., & Taglialatela-Scafati, O. (2009). Marine antimalarials. *Marine Drugs,* 7(2), 130-152.

Faulkner D. J. 1991 Marine natural products. *Nat Prod Rep* 8:97-147, and earlier reports Scheuer P. J. (ed.) 1991 *Bioorganic marine chemistry,* vol IV. Springer-Verlag, Berlin, and earlier volumes.

Fenical, W. (1999). Marine-derived pharmaceuticals and related bioactive agents. *From Monsoons to Microbes, Understanding the Ocean's Role in Human Health,* 73-82.

Fiore, C., Baraghini, A., Shemchuk, O., Sambri, V., Morotti, M., Grepioni, F., & Braga, D. (2022). Inhibition of the Antibiotic Activity of Cephalosporines by Co-Crystallization with Thymol. *Crystal Growth & Design,* 22(2), 1467-1475.

Fouad, M. A., Debbab, A., Wray, V., Müller, W. E. G., & Proksch, P. (2012). New bioactive alkaloids from the marine sponge Stylissa sp. *Tetrahedron,* 68(49), 10176–10179.

Hafez Ghoran, S., & Kijjoa, A. (2021). Marine-Derived Compounds with Anti-Alzheimer's Disease Activities. *Marine Drugs,* 19(8), 410.

Janovská, P., Normant, E., Miskin, H., & Bryja, V. (2020). Targeting casein kinase 1 (CK1) in hematological cancers. *International Journal of Molecular Sciences,* 21(23), 9026.

Kim, S. K., & Karadeniz, F. (2011). Anti-HIV activity of extracts and compounds from marine algae. *Advances in food and nutrition research,* 64, 255-265.

Kiuru P, D'Auria MV, Muller CD, Tammela P, Vuorela H, Yli-Kauhaluoma J. Exploring marine resources for bioactive compounds. *Planta Med.* 2014;80:1234–1246. Marine sponges.

Kobayashi, J., Ohizumi, Y., Nakamura, H., Hirata, Y., Wakamatsu, K. and Miyazawa, T. (1986). Hymenin, a novel a-adrenoceptor blocking agent from the Okinawan marine sponge Hymeniacidon sp. *Experientia,* 42(9), 1064-1065.

Mapelli, M., & Musacchio, A. (2003). The structural perspective on CDK5. *Neurosignals,* 12(4-5), 164-172.

Martín Moyano, P., Němec, V., & Paruch, K. (2020). Cdc-like kinases (CLKs): biology, chemical probes, and therapeutic potential. *International Journal of Molecular Sciences,* 21(20), 7549.

Masch, A., & Kunick, C. (2015). Selective inhibitors of Plasmodium falciparum glycogen synthase-3 (PfGSK-3): new antimalarial agents?. *Biochimica et Biophysica Acta (BBA)-Proteins and Proteomics,* 1854(10), 1644-1649.

Mayer, A. M., & Gustafson, K. R. (2004). Marine pharmacology in 2001–2: antitumour and cytotoxic compounds. *European Journal of Cancer,* 40(18), 2676-2704.

Mohammed, R., Peng, J., Kelly, M., & Hamann, M. T. (2006). Cyclic Heptapeptides from the Jamaican Sponge Stylissa caribica. *Journal of Natural Products,* 69(12), 1739–1744.

Newbold, R. W., Jensen, P. R., Fenical, W., & Pawlik, J. R. (1999). Antimicrobial activity of Caribbean sponge extracts. *Aquatic microbial ecology,* 19(3), 279-284.

Nguyen, T., & Tepe, J. (2009). Preparation of Hymenialdisine, Analogues and Their Evaluation as Kinase Inhibitors. *Current Medicinal Chemistry,* 16(24), 3122–3143.

Rao, K. V., Donia, M. S., Peng, J., Garcia-Palomero, E., Alonso, D., Martinez, A., ... & Hamann, M. T. (2006). Manzamine B and E and ircinal A related alkaloids from an Indonesian Acanthostrongylophora sponge and their activity against infectious, tropical parasitic, and Alzheimer's diseases. *Journal of natural products,* 69(7), 1034-1040.

Renard, E., Gazave, E., Fierro-Constain, L., Schenkelaars, Q., Ereskovsky, A., Vacelet, J., & Borchiellini, C. (2013). Porifera (sponges): recent knowledge and new perspectives. eLS.

Rinehart, K. L. (2007). Secondary Metabolites from Marine Organisms. *Novartis Foundation Symposia,* 236–254.

Rodríguez, I. I., Rodríguez, A. D., Wang, Y., & Franzblau, S. G. (2006). Ileabethoxazole: a novel benzoxazole alkaloid with antimycobacterial activity. *Tetrahedron Letters,* 47(19), 3229-3232.

Scala, F., Fattorusso, E., Menna, M., Taglialatela-Scafati, O., Tierney, M., Kaiser, M., & Tasdemir, D. (2010). Bromopyrrole Alkaloids as Lead Compounds against Protozoan Parasites. *Marine Drugs,* 8(7), 2162–2174.

Semwal, D. K., Kumar, A., Semwal, R. B., & Andola, H. C. (2021). Convolvulus prostratus. In : *Naturally Occurring Chemicals Against Alzheimer's Disease* (pp. 409-424). Academic Press.

Shah B. N. and Seth A. K. (2014). *Textbook of pharmacognosy and phytochemistry.* Elsevier India. 2nd Edition.

Sunada, S., Saito, H., Zhang, D., Xu, Z., & Miki, Y. (2021). CDK1 inhibitor controls G2/M phase transition and reverses DNA damage sensitivity. *Biochemical and Biophysical Research Communications,* 550, 56-61.

Tadesse, S., Caldon, E., Tilley, W., & Wang, S. (2018). Cyclin Dependent Kinase 2 Inhibitors in Cancer Therapy: an Update. *Journal of Medicinal Chemistry.*

Umumararungu, T., Mukazayire, M. J., Mpenda, M., Mukanyangezi, M. F., Nkuranga, J. B., Mukiza, J., & Olawode, E. O. (2020). A review of recent advances in anti-tubercular drug development. *Indian journal of tuberculosis,* 67(4), 539-559.

Vitale, R. M., Rispoli, V., Desiderio, D., Sgammato, R., Thellung, S., Canale, C., Vassalli, M., Carbone, M., Ciavatta, M. L., Mollo, E., Felicità, V., Arcone, R., Gavagnin Capoggiani, M., Masullo, M., Florio, T., & Amodeo, P. (2018). In Silico identification and experimental validation of novel anti-alzheimer's multitargeted ligands from a marine source featuring a "2-aminoimidazole plus aromatic group" scaffold. *ACS Chemical Neuroscience,* 9(6), 1290-1303.

Wang, C., Lu, Y., & Cao, S. (2020). Antimicrobial compounds from marine actinomycetes. *Archives of pharmacal research,* 43(7), 677-704.

World Health Organization. (2022). Retrieved on 03 September from https://www.who.int/news-room/fact-sheets/detail/malaria https://www.who.int/news-room/fact-sheets/detail/hiv-aids

Xu, Y., Yakushijin, K., & Horne, D. A. (1996). Transbromination of brominated pyrrole and imidazole derivatives: Synthesis of the C11N5 marine alkaloid stevensine. *Tetrahedron Letters,* 37(45), 8121–8124.

Xu, Y., Yakushijin, K., & Horne, D. A. (1997). Synthesis of C11N5Marine Sponge Alkaloids: (±)-Hymenin, Stevensine, Hymenialdisine, and Debromohymenialdisine. *The Journal of Organic Chemistry,* 62(3), 456–464.

Yoon, H. R., Balupuri, A., Choi, K. E., & Kang, N. S. (2020). Small molecule inhibitors of DYRK1A identified by computational and experimental Approaches. *International journal of molecular sciences,* 21(18), 6826.

Zhang, H., Khalil, Z., Conte, M. M., Plisson, F., & Capon, R. J. (2012). A search for kinase inhibitors and antibacterial agents: bromopyrrolo-2-aminoimidazoles from a deep-water Great Australian Bight sponge, Axinella sp. *Tetrahedron Letters,* 53(29), 3784–3787.

Zhang, M., Zhang, L., Hei, R., Li, X., Cai, H., Wu, X., ... & Cai, C. (2021). CDK inhibitors in cancer therapy, an overview of recent development. *American journal of cancer research,* 11(5), 1913.

Chapter 7

Bioactive Alkaloids from *Tinospora cordifolia*

Neetika Sharma and Debabrata Sircar*
Department of Biosciences and Bioengineering, Indian Institute of Technology Roorkee, Roorkee, India

Abstract

Plants are a rich source of several bioactive metabolites which finds multi-purpose use in the prevention and cure of human diseases. Bioactive natural products with medicinal value are gradually gaining importance in clinical research due to their well-known property of affordable cost and fewer side effects as compared to drugs. *Tinospora cordifolia* popularly known as named as "Guduchi" or "Giloy" is well known for its efficacy in treating various human diseases in the traditional medicine system. The biological activity of this plant extract comes from its chemical constituents including Alkaloids, Phenolics, Terpenes, Steroids and Lignans. Alkaloids are the predominant bioactive constituents of this plant. In this chapter, we focus on alkaloids derived from the *Tinospora cordifolia*, their structure and function and biosynthesis. Our present study in this chapter encompasses (i) the importance of herbal medicine (ii) general introduction of alkaloids, (iii) major alkaloids reported from *Tinospora cordifolia* and their biological role in disease targeting (iv) large-scale production of alkaloids using in vitro cultures and elicitor-treatment. The conclusion and future scope of the chapter remain in exploiting the metabolic engineering strategy for enhanced alkaloid production from *Tinospora cordifolia*.

Keywords: medicinal plants, cell culture, plant-derived drugs, traditional medicine, Giloy

---

* Corresponding Author Email: debabrata.sircar@bt.iitr.ac.in.

In: The Essential Guide to Alkaloids
Editor: Deepak Kumar Semwal
ISBN: 979-8-88697-456-0

1. Introduction

Plants are rich source of several biologically active natural products. These natural products based on crude plant extracts and their combinations are the major therapeutic back-bone in traditional systems of medicine and treatment. Natural product rich plant-extracts have been used for decades in the traditional medicine system. However, the use of pure isolated and characterized natural products in modern drug discovery and development systems started in the 19th century. Despite of the well-known fact, that plant natural products can cure several diseases efficiently and safely without imposing many side effects, even though they are not very popular as compared to synthetic drugs, due to high isolation cost, low yield, lengthy quality control, and lack of mode of actions. For the past twenty years, there is a growing trend in using plant natural products as drug candidates due to their unique structure and excellent efficacy in curing several deadly diseases (Cragg and Newman, 2005). According to a recent study, in the last 30 years, approximately 50% of approved anticancer drugs are arriving either directly or indirectly from natural products (Newman and Cragg, 2012).

William Withering in the 18th century first time used the extracts of *Digitalis purpurea* to cure oedema, which was the initial step for the systematic clinical study of a medicinal plant to understand its mode of action. Later, Friedrich Sertürner in the 19th century, first time successfully isolated one plant alkaloid, which was probably the landmark for the first rational drug design and discovery from plant sources (Atanasov et al., 2015). Morphine was the first pure plant natural product commercially used as a drug candidate by Merck in 1826. Later on, the first semi-synthetic pure plant-natural product-based drug aspirin (derived from natural product salicin isolated from *Salix alba*), was commercialized by Bayer in 1899 (Veeresham, 2012). This is followed by the isolation of several other plant natural products of medicinal interest such as codeine, quinine, colchicine, atropine, gingerol, digitoxin, *resveratrol* and pilocarpine. Some examples of clinically approved natural product-derived drugs are vincristine, vinblastine, paclitaxel, topotecan, and galantamine as shown in Table 1.

Table 1. List of selected clinically approved natural product-based drugs

S.No.	Generic name	Source plant	Therapeutic action
1	Paclitaxel	*Taxus brevifolia*	Cancer treatment
2	Vincristine	*Catharanthus roseus*	Leukaemia treatment
3	Vinblastine	*Catharanthus roseus*	Lymphoma treatment
4	Artemisinin	*Artemisia annua*	Malaria treatment
5	Colchicine	*Colchicum species*	Gout treatment
6	Galanthamine	*Lycoris radiata*	Alzheimer treatment
7	Ajmalicine	*Rauvolfia sepentina*	Circulatory disorder treatment
8	Atropine	*Atropa belladonna*	Anticholinergic
9	Curcumin	*Curcuma longa*	Anti-amyloidogenic
10	Resveratrol	*Vitis vinifera*	Cognitive impairment
11	Huperzine A	*Huperzia serrata*	Cognitive impairment
12	Morphine	*Opium poppy*	Pain killer

2. Major Classes of Bioactive Plant Secondary Metabolites

Primary metabolites of plants, proteins, nucleic acids, carbohydrates, and lipids, are conserved and essential for normal growth, development and reproduction in plants. However, what makes the plant extract most suitable for disease treatment and cure are the secondary metabolites. Though low in molecular weight and less abundant than primary metabolites, these secondary plant metabolites exert a wide range of biological effects in animals.

These bioactive secondary metabolites now find wide use in pharmaceuticals, cosmetics and even nutraceuticals. Several pharmaceutical activities have been linked to these metabolites, anti-cancer, anti-inflammatory, anti-diabetic, neuro-protective, hepatoprotective, antiviral, antifungal, and antibiotic. Humans have been utilizing the benefits of these metabolites through traditional and herbal medicinal practices. Even now, many modern medicines are based on plant secondary metabolites because of their bioactivity and less toxic nature (Das and Gezici, 2018). Moreover, secondary metabolite-rich plant extracts range from simple to complex structures and from single to compound mixtures. A single molecule may be responsible for specific bioactivity, or a group of bioactive molecules may be particular to the bioactivity. Plant secondary metabolites are classified majorly based on their biosynthesis pathway (González et al., 2019). Three major metabolite groups are phenolics, terpenes and alkaloids (Figure 1).

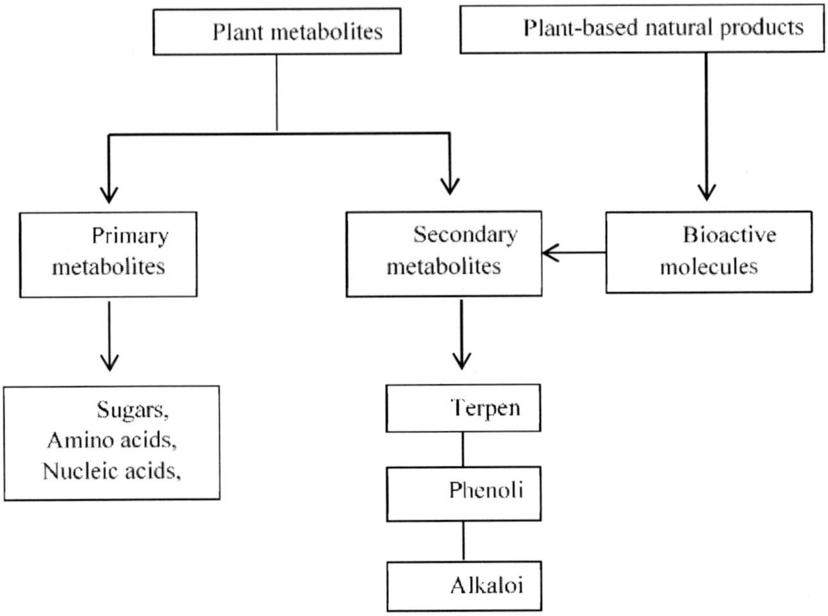

Figure 1. Flowchart showing different types of metabolites present in plants. Natural products derived from plants have various applications because of the bioactive molecules present in plants. Bioactive molecules mainly include the secondary metabolites of plants, the terpenes, phenolics and alkaloids.

2.1. Terpenes

Terpenes constitute a group of secondary metabolites with more than 40,000 different molecules. These are also known as isoprenoids, as the basic structural unit is the isoprene molecule, having five carbons. Terpenes are broadly classified based on the number of the isoprene units, such as hemiterpenes having a single isoprene unit (Fig 2). Many Terpenes are volatile in nature and play a role in pollination and emitting specific odours, as in plants of mint, eucalyptus, ginger and lemon. In central metabolism, these are electron carriers (plastoquinone and ubiquinone), photosynthetic pigments (carotenoids), plant hormones (strigolactones, gibberellins, brassinosteroids) and also participate in protein metabolism. Terpenes are synthesized in plants from primary metabolites from two different pathways: the mevalonic acid pathway (in the cytosol) and the methylerythritol phosphate pathway, MEP (active in the chloroplast) (González et al., 2019).

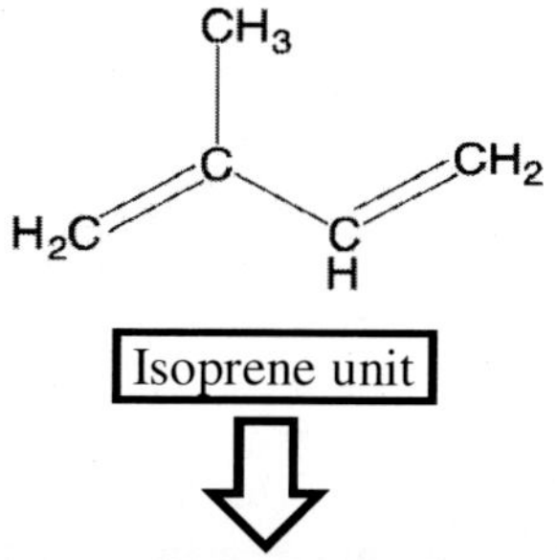

Type of terpene	Number of isoprene unit	Number of carbon atoms present
Hemiterpenes	1	5
Monoterpenes	2	10
Sesquiterpenes	3	15
Diterpenes	4	20
Triterpenes	6	30
Tetraterpenes	8	40

Figure 2. Major categories of bioactive terpenes according to the number of isoprene units present in their structure.

2.2. Phenolics

Phenolics constitute the most abundant plant secondary metabolites. Phenol itself is a benzene moiety having a hydroxyl group. As the name indicates, phenolics are compounds containing one or more phenolic rings where one or more hydroxyl groups are added. These range from derivatives of simple phenolics like caffeic acid and ferulic acid to highly complex and branched phenolic compounds like lignin. Phenolics are further grouped into several classes based on the number of carbon atoms present in a molecule, like phenolic acids, flavonoids, stilbenes, lignans, and tannins. Phenolics are synthesized by the shikimic acid pathway or the malonate/acetate pathway (González et al., 2019). The shikimic acid pathway acts as an essential link between plant primary and secondary metabolism. Through the shikimic acid pathway, aromatic amino acids, namely, L-phenylalanine, L-tryptophan and L-tyrosine, are produced. These amino acids in plants are also precursors of secondary metabolites. L-phenylalanine and L- tyrosine are precursor molecules for phenolic acids, coumarins, chalcones, lignans, cinnamic acids and esters and other derivatives of polyphenols. In the case of flavonoids, the phenylpropanoid ring is formed through the shikimate pathway, and the acetate/mevalonate pathway contributes to the other ring (Figure 3). Due to the presence of hydroxyl groups in these compounds and their structural diversity, the role of polyphenolics in the life of plants is quite significant.

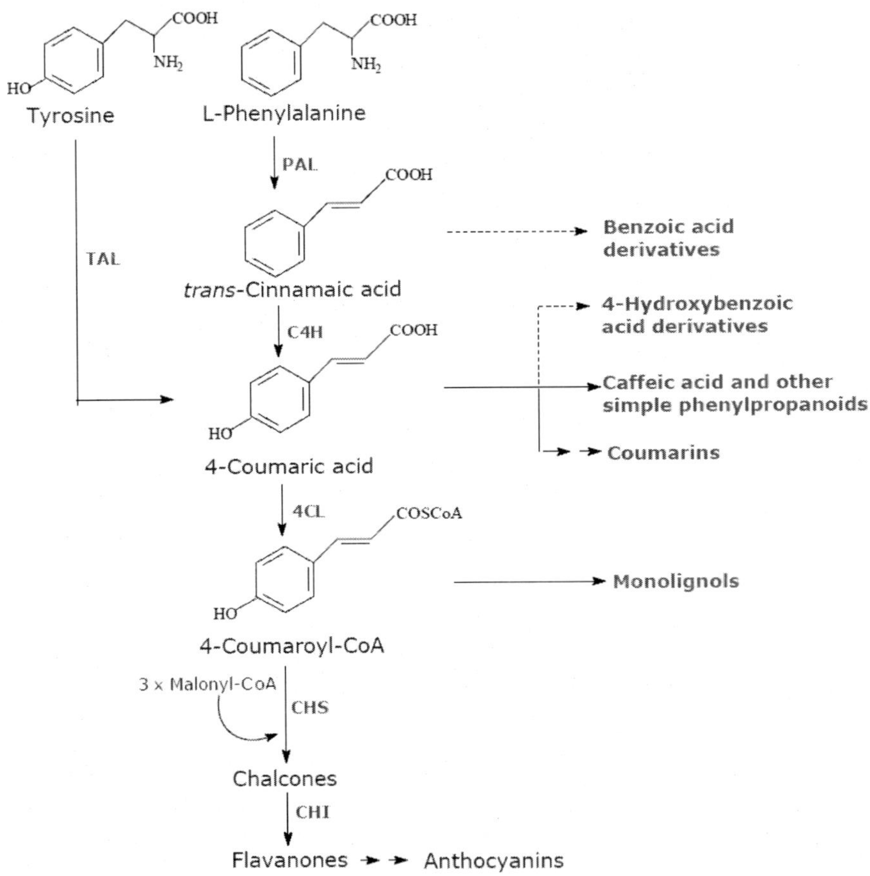

Figure 3. Phenolic biosynthesis pathway leading toward major classes of phenolic metabolites (benzoates, hydroxybenzoates, flavonoids and lignins). TAL, tyrosine ammonia-lyase; PAL, phenylalanine ammonia-lyase; C4H, cinnamate-4-hydroxylase; 4CL, 4-coumarate-CoA-ligase; CHS, chalcone synthase; CHI, chalcone isomerase. Dashed arrow indicates a probable biosynthesis route.

2.3. Alkaloids

Alkaloids are an essential group of structurally diverse compounds that are found in vascular plants and have a nitrogen atom in the core of the molecule. Plants produce a mixture of alkaloids, of which an important constituent dominates. These are considered nature's blessing and curse as they have both beneficial and harmful effects. Typical examples are cinchona bark, used in traditional medicines and ergot poison. Alkaloids are classified on the basis of the precursor molecules as well as the position of the N-atom in the heterocyclic ring. Based on the biogenesis, these are further grouped into true alkaloids, proto alkaloids and pseudo alkaloids as shown in Table 2 (Dey et al., 2020).

Table 2. Classification of major alkaloid classes based on their ring structure and their medicinal uses

Alkaloid type	Ring structure	Alkaloid type	Ring structure
Pyrrole	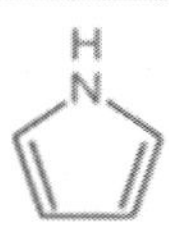	Quinoline	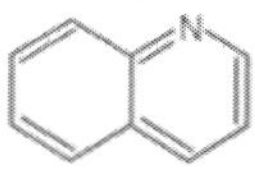
Pyrrolidine	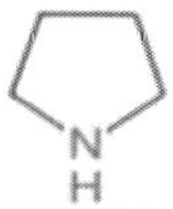	Isoquinoline	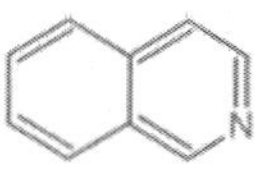
Piperidine	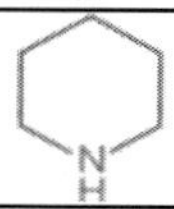	Indole	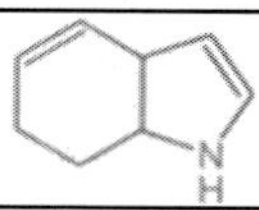
Pyrrolizidine	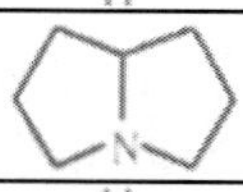	Tropane	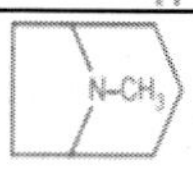
Imidazole		Quinazoline	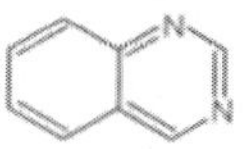

2.3.1. True Alkaloids

These alkaloids contain nitrogen in the heterocyclic ring system. Their precursors are amino acids like L-phenylalanine/ L-tyrosine, L-lysine, L-ornithine, and L- Histidine. All true alkaloids are bitter in taste and solid, except nicotine which is a liquid. These alkaloids occur in three states: free molecules, as N-oxides and as salts.

2.3.2. Proto Alkaloids

These alkaloids do not have nitrogen as a part of the heterocyclic ring system, though the source of the amino group is an amino acid. L-tryptophan and L-tyrosine are the precursor molecules for proto alkaloids. Yohimbine, mescaline and hordenine are some examples of proto alkaloids.

2.3.3. Pseudo Alkaloids

These alkaloids do not have the basic carbon skeleton derived from amino acids. These alkaloids are derived by pathways connected to amino acid pathways by deamination and transamination with intermediates of the pathways. Pseudo-alkaloids can also be produced from non-amino acid precursors. Common examples are aconite, theobromine and coniine.

2.3.4. Classification of Alkaloids

Based on ring structure, the alkaloids are classified into ten major classes as shown in Table 3. The table shows some examples of alkaloid classes based on their biosynthetic origin.

Table 3. Examples of alkaloid classes based on amino-acid precursors used in their biosynthesis

Type of alkaloid	Class of alkaloid	Examples of chemical group
Tryptophan-derived alkaloids	True alkaloid	Indole alkaloids Quinoline alkaloids Ergot alkaloids
	Proto alkaloids	Terpenoid indole alkaloids
Ornithine-derived alkaloids	True alkaloid	Pyrrolizidine alkaloids Tropane alkaloids Pyrrolidine alkaloids
Histidine-derived alkaloids	True alkaloid	Imidazole alkaloids
Nicotinic acid-derived alkaloids	True alkaloid	Pyridine alkaloids
Lysine-derived alkaloids	True alkaloid	Indolizidine alkaloids Quinolizidine alkaloids Piperidine alkaloids
Anthranilic acid-derived alkaloids	True alkaloid	Quinoline alkaloids
Tryptophan-derived alkaloids	Proto alkaloids	Terpenoid indole alkaloids

2.3.5. Distribution and Biological Activity of Alkaloids

Alkaloids differ in their distribution within the same plant as well as different plant species. Some families are rich sources of alkaloids like Papaveraceae, Ranunculaceae, Solanaceae and Amaryllidaceae. Multiple functions have been attributed to alkaloids related to self-preservation, communication and inhibition of competitors. Alkaloids readily form H-bonds with proteins, enzymes and may receptors because of the presence of protons accepting nitrogen atoms and one or several protons donating amine hydrogen atoms. Moreover, the presence of other electron donating and accepting functional groups such as hydroxyl and phenolic moieties also explain the bioactivity of alkaloids. Pharmacological properties include central nervous depressant (e.g., morphine), central nervous stimulant (e.g., brucine), analgesic (e.g., codeine), antitumor (e.g., vinblastine), antihyperten-sive (e.g., reserpine), antihypotensive (e.g., ephedrine), anticholinergic (e.g., atropine), antiemetic (e.g., scopolamine) antipyretic (e.g., quinine), vaso-constrictor and oxytocic (e.g., ergometrine), and antimalarial (e.g., quinine) activities. These activities are exploited both in traditional medicine (e.g., quinine-rich cinchona bark in the treatment of malaria) as well as in modern medicine. Some plant alkaloids are highly toxic. In studies of 350 plant pyrrolizidine alkaloids, approximately 50% were hepatotoxic and several carcinogenic (Stegelmeier et al., 1999). The well-known historical examples of alkaloid toxicity include belladonna, used as a poison in Roman times, and ergot, responsible for thousands of deaths in the Middle Ages (Cushnie et al., 2014).

3. Bioactive Phytoconstituents from *Tinospora cordifolia*

One of the important medicinal plants used in the Indian ayurvedic system and traditional practices is *Tinospora cordifolia*, commonly known as "Guduchi" or "Giloy". This plant is a large deciduous climbing shrub belonging to the family, Menispermaceae. The stem of this plant is filiform, woody and fleshy, the leaf is heart-shaped, flowers are unisexual and fruits are single seed-bearing. The plant occurs in diverse morphological forms (Sharma et al., 2019) as shown in Figure 4. The Sanskrit term "Guduchi" means one which protects the body. Another name for this plant is "Giloy," meaning heavenly elixir, attributed to its ability to impart vitality and longevity. It is a wild, perennial, glabrous climbing shrub. It has heart-shaped leaves, hence the name "*cordifolia*". The stem of this plant is corky, succulent, grooved with lenticels, creamy white to grey in colour and has aerial roots (Kumar et al., 2017). This plant is extensively used in Ayurveda and traditional medicine system (Meena et al., 2010). The potential medicinal properties of this plant as reported by scientific research include antipyretic, anti-diabetic, antispasmodic, anti-inflammatory, anti-oxidant, anti-arthritic, anti-allergic, anti-leprotic, antimalarial, anti-stress, immuno-modulatory, hepato-protective and anti-neoplastic activities (Modi et al., 2021).

Figure 4. Different plant parts of *Tinospora cordifolia*. a) Whole climbing plant b) leaf c) stem d) root e) flowers.

The wide spectrum of its medicinal properties is attributed to the rich phytochemistry of this plant. *T. cordifolia* is rich in several classes of natural products, including alkaloids, glycosides, diterpenoid lactones, steroids, sesquiterpenoid, aliphatic compounds, phenolics, essential oils (Singh et al., 2017; Sharma et al., 2019). Giloy itself has a bitter, astringent and pungent taste because of its constituents. It is believed that these bitter constituents detoxify the tissue by destroying the toxic metabolic products and this bitter taste was the reason for its use by traditional medicinal practitioners to improve metabolic activity. Several studies have been conducted to study the activities of different extracts of Giloy (Kumar et al., 2018; Sharma et al., 2012; Upadhyay et al., 2010). Among these phytoconstituents of *T. cordifolia*, alkaloids play a key role in exhibiting the biological activity of this plant. This chapter will deal with the primary alkaloids present in *T. cordifolia*, their chemical nature, and the bioactivity studied so far.

3.1. Bioactive Alkaloids from Tinospora cordifolia

Tinospora cordifolia is rich in several types of phytoconstituents, but its alkaloidal fraction has been extensively screened for various biological activities. Interestingly, several alkaloids reported from this plant is also present in other plant sources belonging to the family Berberidaceae and Ranunculaceae (Rolle et al., 2021). Twelve bioactive alkaloids from this plant have been extensively studied as shown in Table 4 and Figure 5.

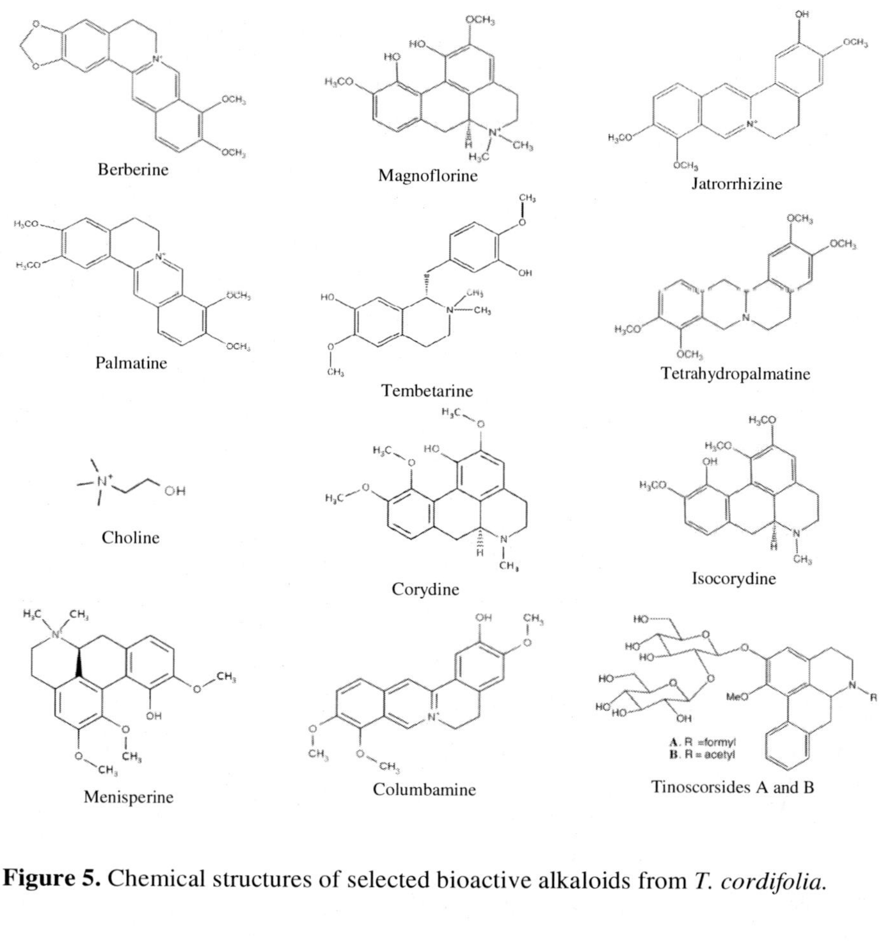

Figure 5. Chemical structures of selected bioactive alkaloids from *T. cordifolia*.

Table 4. Major bioactive alkaloids reported from *Tinospora cordifolia*

S.No.	Name of alkaloid	Bioactivity	Reference
1	Berberine	anticancer, antioxidant, anti-inflammatory, antidiabetic, anti-obesity, neuroprotective, nephroprotective, cardioprotective, hepatoprotective, cholesterol-lowering	Neag et al.,2018; Liu et al.,2019; Li et al., 2014; Ahmed et al., 2015; Wang et al., 2015; Och et al,2020; Domitrovi´c et al.,2011
2	Magnoflorine	anti-Alzheimer's, anticancer, antioxidant, anti-inflammatory, antidiabetic, immunomodulatory, anti-allergic, anti-depressant, anti-osteoporotic, antibacterial, antifungal, antiviral, anti-ageing	Okoń et al., 2020; Baran et al., 2020; Guo et al., 2018; Sun et al., 2020; Wei et al., 2020;
3	Jatrorrhizine	anti-rheumatoid arthritis, anti-Alzheimer's, anti-inflammatory, anti-cancer, anti-microbial, anti-obesity, anti-diabetic, anti-protozoal, anti-depressant, neuroprotective, hepatoprotective, bone-protective, hypolipidemic	Zhong et al., 2022; Rolle et al., 2021; Patel et al., 2012; Liu et al, 2020; Ali et al., 2013 Yu et al., 2019;
4	Palmatine	anticancer, antioxidant, anti-inflammatory, antibacterial, antiviral, anti-protozoal, gastroprotective, hepatoprotective, anti-obesity, antidiabetic, hypolipidemic, neuroprotective	Ekeuku et al., 2020; Kim et al., 2009; Hung et al., 2008; Jung et al., 2009; Tsai and Lee,2 010; Dhingra and Bhankher, 2014; Shen et al.,2018; Long et al., 2019;
5	Tembetarine	Anti-inflammatory, Antinociceptive, Anti-diabetic	Cordeiro et al., 2016; Annam et al., 2022
6	Tetrahydropalmatine	analgesic, neuroprotective, anti-addictive, anti-inflammatory, anti-cancer, hypothermic	Lin et al., 2001; Li et al., 2021; Du et al., 2022
7	Choline	anti-oxidant, neuroprotective, combat toxicity, important for the development of neural development of the fetus, membrane function and neurotransmission	Biswas & Giri, 2015; Arias et al., 2020
8	Corydine	antinociceptive, anti-protozoal, DNA damaging	Semwal et al., 2010; Kaserer et al., 2020
9	Isocorydine	anticancer	Zhong et al., 2014
10	Menisperine	anti-inflammatory	Wang et al., 2019
11	Columbamine	suppresses hepatocellular carcinoma	Lin et al.,2019
12	Tinoscorsides (A and B)	not well defined	Phan et al., 2010

A few important alkaloids of *T. cordifolia* and their major biological activity have been briefly described below.

3.1.1. Berberine

Berberine is a quaternary benzylisoquinoline plant alkaloid that occurs as an active constituent in *T. cordifolia*. This alkaloid is also reported from several other plants belonging to the families Berberidacaceae, Annonaceae, Menispermaceae, Papaveraceae, Ranunculaceae and Rutaceae. Several pharmacological effects have been attributed to this alkaloid and the plants containing it. Traditional and modern medicines have emphasized that berberine has various pharmacological activities, such as analgesic, hypolipidemic, antimicrobial, anti-inflammatory, blood-pressure lowering and inhibitory to various cancers, including lung cancer, ovarian cancer, liver cancer and cervical cancer (Liu et al., 2019). Some of its activities are discussed briefly.

3.1.1.1. Anti-Oxidant Activity

During normal conditions, a balance is maintained between anti and prooxidant agents (ROS, RNS) in the body. The imbalance between these two can lead to oxidative stress. Berberine shows partial anti-oxidant properties due to its reverse effect on lipid peroxidation after carcinogenesis induction was done (Li et al., 2014). Other mechanisms involved with the anti-oxidant role of berberine are free oxygen removal, ROS/RNS scavenging, oxidation/transformation of certain substances or enhancing the anti-oxidant effect of some endogenous substances (Neag et al., 2018). The antioxidant activity of berberine also explains renal function improvement in the case of diabetic nephropathy (Liu et al., 2016).

3.1.1.2. Anti-Cancer Activity

Cancer development can be attributed to several factors, cell cycle altercations, irregulated apoptosis and autophagy. Studies showed that berberine regulates the cell cycle and inhibits cell proliferation by affecting cell cycle progression (Liu et al., 2019). In addition, studies also showed that berberine regulates apoptosis by promoting the expression of caspase-8, -9, and caspase-3, which inhibits the expression of bcl-2, and Bim and Bax. With respect to autophagy, berberine shows dual effects, inhibition as well as activation. In several studies, berberine inhibits cell proliferation by inducing autophagy and also reverses drug resistance by regulating cell autophagy

(Wang et al., 2015). However, in mature adipocytes, berberine maintained cellular homeostasis by inhibiting autophagy (Wang et al., 2020). In breast cancer cells, berberine inhibited cell proliferation and metastasis by downregulation of TGF-β1 and targeting ephrin-B2. In all, berberine inhibits cell invasion and metastasis by affecting the expression of tumour-related signaling pathways and proteins. Berberine also regulates the tumour microenvironment by affecting inflammatory and immune molecules. Berberine, in osteosarcoma, altered the inflammatory response by downregulating the IL-1β /caspase-1 signaling pathway, which resulted in cell apoptosis. Though BBR exerts beneficial effects that may aid in the treatment of tumours, the efficacy of berberine is limited by poor solubility in water, the low absorption rate in intestines and rapid metabolism. Therefore, new formulations need to be developed that improve the absorption of berberine in the intestines, which may have great potential for the treatment of cancer (Wang et al., 2020).

3.1.1.3. Hepatoprotective and Neuroprotective Activities

The hepatoprotective nature of berberine was studied y induce hepatotoxicity in lab rats using doxorubicin. Rats that were given berberine pretreatment showed reduced histological damage. Berberine works by lowering oxidative and nitrosamine-caused stress and modulating the inflammatory response that leads to favourable changes in the liver. Increased SOD activity and decreased lipid peroxidation contribute to reduced levels of COX-2, TNF-α, and induced nitric oxide species. Moreover, the decrease in transaminase level also supports the hypothesis that berberine aids in maintaining the integrity of the hepatocellular membrane (Domitrović et al., 2011). Hyperlipidemia and hyperglycemic abilities of berberine strongly support their protective role in neurological disorders. Recent studies have supported the ability of berberine to protect the CNS, which makes it a potent agent in disorders like anxiety, Alzheimer's disease, cerebral ischemia, schizophrenia and mental depression (Och et al., 2020).

3.1.2. Magnoflorine

Magnoflorine is an important alkaloid with an aporphine nucleus. It is a type of quaternary aporphine alkaloid, a subclass of quinolone alkaloids. It is a derivative of (S)-corytuberine, which has been quaternized by the addition of another methyl group to the nitrogen atom. This alkaloid is highly polar and water soluble. This alkaloid is reported to be present in *T. cordifolia* and in several other plants belonging to the family Menispermaceae, Rutaceae,

Ranunculaceae, Magnoliaceae, and Berberidaceae. In recent years, much research has been undertaken to review the pharmacological activities of magnoflorine. The results have indicated that magnoflorine exhibits a broad range of pharmacological properties, anxiolytic, anti-inflammatory, antifungal, anti-diabetic, anti-oxidant, anti-cancer, and anti-hypertensive (Baran et al., 2020).

3.1.2.1. Anti-Inflammatory Activity

Inflammation is a biological response to extreme challenges to the body's homeostasis, such as stress, injury, or infection. Inflammation can be acute or chronic. It has been shown that magnoflorine, at higher concentrations, exhibited anti-inflammatory activity in the case of acute lung injury (Guo et al., 2018). Mechanistically, magnoflorine works by reducing the expression of pro-inflammatory molecules TNF-a, IL-1b, and IL-6 and inhibiting TLR4-mediated NF-kB and MAPK signaling pathways.

3.1.2.2. Antifungal Activity

Fungal infections are of major concern for human health. A study was conducted to evaluate the antifungal activity of 44 compounds against candida strains (Kim et al., 2018). The highest growth inhibitory activity against the tested strains was shown by magnoflorine. The antifungal activity of magnoflorine against *Trichophyton rubrum,* the causal organism of the Athlete's foot, was studied. Magnoflorine inhibited the hyphal growth along with the conidia germination and affected the mycelial morphology. Moreover, the fungal cell membrane of *T. rubrum* was destroyed by increased nucleic acid leakage, reduced squalene oxidase, CYP51 enzyme activities, and decreased ergosterol content in hyphae. All these studies suggested the therapeutic application of magnoflorine as a potent antifungal agent (Luo et al., 2021).

3.1.2.3. Anticancer Activity

Autophagy has dual roles; it can be "pro-survival" or "pro-death" in response to different biological conditions. The role of magnoflorine in inhibiting human gastric cancer progression was studied by Sun et al. (2020). It was observed that magnoflorine induced autophagic cell death in gastric cancer cell lines and prevented gastric cancer development. In cancer, AKT/mTOR signaling pathways display a pivotal role in the cell death of diverse tumour cells by autophagy. Magnoflorine treatment induced autophagy by decreased expression of p-AKT/p-mTOR, leading to suppression of the AKT/mTOR pathway. It was concluded that magnoflorine could significantly reduce tumour growth and weight without minor damage to liver, kidney and heart issues in a laboratory animal (Wei et al., 2020). One of the severe issues faced in cancer treatments is the toxicity of a molecule in the long term. Doxorubicin (DOX) is a chemotherapeutic agent used in combination with other medicines to treat various types of cancers. Magnoflorine was suggested as a promising molecule for application in combination with DOX to reduce the toxic effects of DOX (Wei et al., 2020). Isobolographic analysis was done to study the synergistic effect of both these compounds, and the results indicated that Magnoflorine enhanced the sensitivity and DNA damage to breast cancer cells to DOX treatment. In another study, Okoń et al. (2020) investigated the influence of magnoflorine on proliferation, viability, cell cycle arrest and induction of apoptosis in several cell lines, namely T98G glioma, MDA-MB-468 breast, NCI-H1299 lung and TE671 rhabdomyosarcoma cancer cells. It was observed that magnoflorine significantly reduced the cell viability, slightly reduced proliferation, induced apoptosis and influenced cell cycle progression in the S/G2 phase of the studied cancer lines.

3.1.3. Palmatine

Palmatine is a proto-berberine isoquinoline alkaloid that exists as a quaternary ammonium salt. It is one of the major bioactive compounds of *T. cordifolia*. This alkaloid is also present in *Berberis* spp., *Papaver* spp., *Enatia* spp., *Coptis* spp. and *Tinospora* spp., belonging to families Berberidaceae, Papaveraceae, Annonaceae, Ranunculaceae and Menisper-maceae. The scientific literature describes the occurrence of palmatine together with berberine in many families. Due to the higher content of berberine in the plant extracts, palmatine is often less isolated and tested. However, as per the Ayurvedic and traditional medicinal knowledge, palmatine has been attributed as a remedy against jaundice, liver-related disorders, inflammation, dysentery and hypertension (Tarabasz and Kukuła-Koch., 2020).

3.1.3.1. Anti-Oxidant and Anti-Inflammatory Activities

Metabolic disorders have been linked with oxidative stress, an imbalance between oxidants and antioxidants. Palatine, in both in-vivo and in-vitro systems, has shown antiadipogenic activity. A recent study with 4-week-old Syrian golden hamsters fed with a high-fat diet showed reduced body weight and total serum cholesterol after being treated with palmatine for 4 weeks (Ekeuku et al., 2020). One of the leading causes of death in developing countries is myocardial infarction; also, a risk factor contributed by metabolic disorders. Moreover, pro-oxidative and pro-inflammatory conditions in metabolic disorders could damage the walls of blood vessels or trigger coronary microvascular dysfunction. The cardioprotective activity of palmatine has also been reported in various studies (Kim et al., 2009).

3.1.3.2. Neuroprotective Activity

Alzheimer's disease is a neurodegenerative disease that gradually develops with age and leads to a decline in cognition. More than 50 million people in the world suffer from Alzheimer's (Zhang et al., 2021). Abnormalities specific to this disease include neuronal damage, synaptic changes and brain inflammation. Accumulation of extracellular amyloid and sere synaptic connection dysfunction and neuronal death lead to brain atrophy. Studies have shown that palmatine inhibits the activity of acetylcholine esterase, neuro transaminase and butyrylcholine esterase and thus acts as anticholinergic drugs that have been the target of many studies for the treatment of this disease (Tsai and Lee, 2010). Palmatine also reduces the formation of ROS and RNS, which may cause inflammation. Research shows that palmatine exhibits an antidepressant effect by regulating monoamine oxidase-A (MAO-A) activity, brain catalase levels, lipid peroxidation, corticosterone and plasma nitrite levels (Dhingra and Bhankher, 2014).

3.1.3.3. Anti-Cancer Activity

Palmatine exhibit anticancer activity by binding with single-stranded RNA or double stranded DNA. Palmatine also works photo toxically. It produces O_2 after binding with DNA, which causes photooxidation of guanine and leads to DNA break or cleavage (Hirakawa et al., 2005). The ability of palmatine to bind to various nucleic acids attributes to its anti-cancer activity. Palmatine had growth inhibitory effects on several human cancer cell lines: 7701QGY, HepG2, SMMC7721, CEM, K III, CEM/VCR. In addition, palmatine also exhibited anti-cancer activity on U251, CHOK-1, KB, MCF-7, HT-29 and SiHa cell lines (Bala et al., 2015).

3.1.4. Tembetarine

Tembetarine is a benzylisoquinoline alkaloid that is formed by the methylation of the tertiary amino group of reticuline. It is present naturally in *T. cordifolia.* Though no individual studies on this compound are available in the literature, various experiments and in silico studies have been done using the plant extracts of these medicinally important plants. Extracts of this plant is used for their anti-inflammatory and antinociceptive activities. An in-silico study on *T. cordifolia* and *Curcuma longa* for pulmonary fibrosis inhibitors showed that their anti-inflammatory properties could aid in anti-fibrotic drug discovery. Magnoflorine and tembetarine were selected as potent inhibitors of TGF-βRI because of their high-affinity binding to the active site of the main protein, TGF-βRI. In another network pharmacology-based assessment to identify the anti-diabetic action of *T. cordifolia*, nine phytoconstituents were selected. Of these nine, tembetarine showed the highest drug-likeness hit and had maximum interaction with receptor proteins related to diabetes. However, these results are based on simulations, and further research needs to be undertaken to validate these findings (Annam et al., 2022).

3.1.5. Jatrorrhizine

Jatrorrhizine is a known isoquinoline alkaloid of the protoberberine type. This metabolite is found in various plant families, Berberidaceae, Menispermaceae, Papaveraceae, Ranunculaceae, Annonaceae, and Rutaceae (Zhong et al., 2022). Many plants from these families are extensively used in folk medicines and traditional medicinal systems, like species of *Tinospora, Mahonia, Berberi* and *Coptis* (Zhong et al., 2022). The plant extracts are known to be antipyretic, analgesic, and anti-inflammatory to treat abdominal pain, diarrhoea and gastrointestinal disorders (Rolle et al., 2021).

3.1.5.1. Anti-Diabetic Activity

Jatrorrhizine inhibits the activity of aldose reductase, and α-glucosidase and thus, acts as a hypoglycaemic agent (Patel et al., 2012). Anti-diabetic activity of this alkaloid in in-vivo (glucose-loaded rats and hyperlipidemic mice) and in-vitro systems (RINm5F cells and HepG2 cells) are displayed via the promotion of insulin secretion, enhancing glucose tolerance and insulin sensitivity and inhibition of hepatic gluconeogenesis.

3.1.5.2. Anti-Microbial and Anti-Protozoal Activities

Jarorrhizine strongly inhibited the growth of some bacteria, namely *Candida auris*, *Candida albicans, Propionibacterium acnes*, *Candida tropicalis, Phylococcus aureus*, and *Staphylococci*. The basis for the antibacterial effect was the expression of the NorA multi-drug efflux pump and inhibition of drug efflux (Yu et al., 2019). In-vitro antiprotozoal research on jatrorrhizine has shown its anti-plasmodial activity against several drug-resistant strains of *Plasmodium falciparum.* Anti-trypanosomal activity against the *Trypano-*

soma brucei rhodesiense STIB and anti-leishmanial activity against *Leishmania donovani axenic* MHOM-ET-67/82 strain.

3.2. Enhanced and Targeted Production of Bioactive Alkaloids from *T. cordifolia*

Moreover, plants show variations in their biochemical profiles due to geography, season, time of harvest and also storage. This variability can significantly affect the pharmacological activity of the final product even if the product is prepared using the same amount of raw material. The ultimate source of traditional medicines, herbal remedies and natural products are the plants growing in the wild. The growing demand for herbal products in the market could drive the over-harvesting of plants in the wild and threaten biodiversity. Poor cultivation and collection practices can lead to the extinction of threatened and endangered plant species and cause irreparable loss of natural resources. Also, another issue with herbal medicines is their quality control, which directly impacts their safety and efficacy. Herbal extracts may be adulterer, contaminated or may contain toxic compounds. Research into the safety, quality, molecular effects and clinical efficacy of the commonly used herbs is needed. Biochemical fingerprinting using hyphenated techniques like GC-MS and LC-MS for definitive authentication and quality control of herbal products should be done. These studies will help in identifying the active metabolites in the plants. Moreover, experiments should be conducted to identify the bioactivities of the active metabolites. Studies should also be undertaken to elucidate the biochemical pathway involved in the production of these active compounds. Plant cell culture provides a powerful production system for natural products independent from constraints resulting from geographical and seasonal variations and also limitations resulting from dwindling wild plant populations. Therefore, this platform can help in the establishment of a robust supply chain and help alleviate ecological problems and the loss of biodiversity. Several bioactive alkaloids such as acridines, thebaine, galanthamine, tropane alkaloid, indole alkaloids are being produced by plant cell cultures (Chandran et al., 2020).

Conclusion and Future Scope

Plant secondary metabolites have been an integral part of human civilization for hundreds of years. In recent years, the scientific community has shown immense interest in natural products. Due to the increasing population, and hence the increasing demands for these products, it is essential to find alternate ways for their production. Overexploitation of natural plant sources has already threatened the existence of many species, and many have already become extinct. Future research needs to be undertaken to identify the potent bioactive alkaloids from *T. cordifolia*, their biosynthetic and transport pathways, mechanism of bioactivity and bioefficacy. Without burdening the natural resources, these natural products can be produced using various bioengineering techniques.

References

Ahmed, T., Gilani, A. U., Abdollahi, M., Daglia, M., Nabavi, S. F., & Nabavi, S. M. (2015). Berberine and neurodegeneration: A review of literature. *Pharmacological reports: PR*, *67*(5), 970–979. https://doi.org/10.1016/j.pharep.2015.03.002.

Ali, S., Igoli, J., Clements, C., Semaan, D., Alamzeb, M., Rashid, M. U., Shah, S. Q., Ferro, V. A., Gray, A. I., & Khan, M. R. (2013). Antidiabetic and antimicrobial activities of fractions and compounds isolated from *Berberis brevissima* Jafri and *Berberis parkeriana* Schneid. *Bangladesh Journal of Pharmacology*, *8*(3), 336–342. https://doi.org/10.3329/bjp.v8i3.13888.

Annam, N., Karlapudi, A. P., Doble, M., Srirama, K., & T C, V. (2022). An *in silico* study on pulmonary fibrosis inhibitors from *Tinospora cordifolia* and *Curcuma longa* targeting TGF-β RI. *Journal of biomolecular structure & dynamics*, 1–17. Advance online publication. https://doi.org/10.1080/07391102.2022.2029772.

Arias, N., Arboleya, S., Allison, J., Kaliszewska, A., Higarza, S. G., Gueimonde, M., & Arias, J. L. (2020). The Relationship between Choline Bioavailability from Diet, Intestinal Microbiota Composition, and Its Modulation of Human Diseases. *Nutrients*, *12*(8), 2340. https://doi.org/10.3390/nu12082340.

Atanasov, A. G., Waltenberger, B., Pferschy-Wenzig, E. M., Linder, T., Wawrosch, C., Uhrin, P., Temml, V., Wang, L., Schwaiger, S., Heiss, E. H., Rollinger, J. M., Schuster, D., Breuss, J. M., Bochkov, V., Mihovilovic, M. D., Kopp, B., Bauer, R., Dirsch, V. M., & Stuppner, H. (2015). Discovery and resupply of pharmacologically active plant-derived natural products: A review. *Biotechnology advances*, *33*(8), 1582–1614. https://doi.org/10.1016/j.biotechadv.2015.08.001.

Bala, M., Pratap, K., Verma, P. K., Singh, B., & Padwad, Y. (2015). Validation of ethno-medicinal potential of *Tinospora cordifolia* for anticancer and immunomodulatory activities and quantification of bioactive molecules by HPTLC. *Journal of ethnopharmacology, 175*, 131–137. https://doi.org/10.1016/j.jep.2015.08.001.

Baran, M., Miziak, P., Bonio, K. (2020). Magnoflorine – a compound with anti-tumour activity. *Journal of Pre-Clinical and Clinical Research, 14*(3), 98-101. https://doi.org/10.26444/jpccr/127326.

Biswas, S., & Giri, S. (2015). Importance of choline as essential nutrient and its role in prevention of various toxicities. *Prague medical report, 116*(1), 5–15. https://doi.org/10.14712/23362936.2015.40.

Chandran, H., Meena, M., Barupal, T., & Sharma, K. (2020). Plant tissue culture as a perpetual source for production of industrially important bioactive compounds. *Biotechnology reports (Amsterdam, Netherlands), 26*, e00450. https://doi.org/10.1016/j.btre.2020.e00450.

Cordeiro, K. W., Felipe, J. L., Malange, K. F., do Prado, P. R., de Oliveira Figueiredo, P., Garcez, F. R., de Cássia Freitas, K., Garcez, W. S., & Toffoli-Kadri, M. C. (2016). Anti-inflammatory and antinociceptive activities of *Croton urucurana* Baillon bark. *Journal of ethnopharmacology, 183*, 128–135. https://doi.org/10.1016/j.jep.2016.02.051.

Cragg, G. & Newman, D. (2005). Biodiversity: A continuing source of novel drug leads. *Pure and Applied Chemistry, 77*(1), 7-24. https://doi.org/10.1351/pac200577010007.

Cushnie, T. P., Cushnie, B., & Lamb, A. J. (2014). Alkaloids: an overview of their antibacterial, antibiotic-enhancing and antivirulence activities. *International journal of antimicrobial agents, 44*(5), 377–386. https://doi.org/10.1016/j.ijantimicag.2014.06.001.

Das, K., & Gezici, S. (2018). Secondary plant metabolites, their separation and identification, and role in human disease prevention. *Annals of Phytomedicine: An International Journal, 7*, 13-24. DOI: 10.21276/ap.2018.7.2.3.

Dey, P., Kundu, A., Kumar, A., Gupta, M., Lee, B. M., Bhakta, T., Dash, S., Kim, H. S. (2020). Analysis of alkaloids (indole alkaloids, isoquinoline alkaloids, tropane alkaloids). *Recent Advances in Natural Products Analysis*.505–567. https://doi.org/10.1016/B978-0-12-816455-6.00015-9.

Dhingra, D., & Bhankher, A. (2014). Behavioral and biochemical evidences for antidepressant-like activity of palmatine in mice subjected to chronic unpredictable mild stress. *Pharmacological reports: PR, 66*(1), 1–9. https://doi.org/10.1016/j.pharep.2013.06.001.

Domitrović, R., Jakovac, H., & Blagojević, G. (2011). Hepatoprotective activity of berberine is mediated by inhibition of TNF-α, COX-2, and iNOS expression in CCl(4)-intoxicated mice. *Toxicology, 280*(1-2), 33–43. https://doi.org/10.1016/j.tox.2010.11.005.

Du, Q., Meng, X., & Wang, S. (2022). A Comprehensive Review on the Chemical Properties, Plant Sources, Pharmacological Activities, Pharmacokinetic and Toxicological Characteristics of Tetrahydropalmatine. *Frontiers in pharmacology, 13*, 890078. https://doi.org/10.3389/fphar.2022.890078.

Ekeuku, S. O., Pang, K. L., & Chin, K. Y. (2020). Palmatine as an Agent Against Metabolic Syndrome and Its Related Complications: A Review. *Drug design, development and therapy, 14*, 4963–4974. https://doi.org/10.2147/DDDT.S280520.

González Mera, I. F., González Falconí, D. E., & Morera Córdova, V. (2019). Secondary metabolites in plants: main classes, phytochemical analysis and pharmacological activities. *Bionatura*, *4*, 1000-1009. http://dx.doi.org/10.21931/RB/2019.04.04.11

Guo, S., Jiang, K., Wu, H., Yang, C., Yang, Y., Yang, J., Zhao, G., & Deng, G. (2018). Magnoflorine Ameliorates Lipopolysaccharide-Induced Acute Lung Injury via Suppressing NF-κB and MAPK Activation. *Frontiers in pharmacology*, *9*, 982. https://doi.org/10.3389/fphar.2018.00982

Hirakawa, K., Kawanishi, S., & Hirano, T. (2005). The mechanism of guanine specific photooxidation in the presence of berberine and palmatine: activation of photosensitized singlet oxygen generation through DNA-binding interaction. *Chemical research in toxicology*, *18*(10), 1545–1552. https://doi.org/10.1021/tx0501740

Hung, T. M., Na, M., Dat, N. T., Ngoc, T. M., Youn, U., Kim, H. J., Min, B. S., Lee, J., & Bae, K. (2008). Cholinesterase inhibitory and anti-amnesic activity of alkaloids from *Corydalis turtschaninovii*. *Journal of ethnopharmacology*, *119*(1), 74–80. https://doi.org/10.1016/j.jep.2008.05.041

Jung, H. A., Min, B. S., Yokozawa, T., Lee, J. H., Kim, Y. S., & Choi, J. S. (2009). Anti-Alzheimer and antioxidant activities of Coptidis Rhizoma alkaloids. *Biological & pharmaceutical bulletin*, *32*(8), 1433–1438. https://doi.org/10.1248/bpb.32.1433

Kaserer, T., Steinacher, T., Kainhofer, R., Erli, F., Sturm, S., Waltenberger, B., Schuster, D., & Spetea, M. (2020). Identification and characterization of plant-derived alkaloids, corydine and corydaline, as novel mu opioid receptor agonists. *Scientific reports*, *10*(1), 13804. https://doi.org/10.1038/s41598-020-70493-1

Kaur, R., Arora, S. (2015). Alkaloids-important therapeutic secondary metabolites of plant origin. *Journal of Critical Review*. *2*(3),1–8. DOI: 10.31838/jcr.02.01.01

Kim, J., Ha Quang Bao, T., Shin, Y. K., & Kim, K. Y. (2018). Antifungal activity of magnoflorine against *Candida* strains. *World journal of microbiology & biotechnology*, *34*(11), 167. https://doi.org/10.1007/s11274-018-2549-x

Kim, Y. M., Ha, Y. M., Jin, Y. C., Shi, L. Y., Lee, Y. S., Kim, H. J., Seo, H. G., Choi, J. S., Kim, Y. S., Kang, S. S., Lee, J. H., & Chang, K. C. (2009). Palmatine from Coptidis rhizoma reduces ischemia-reperfusion-mediated acute myocardial injury in the rat. *Food and chemical toxicology: an international journal published for the British Industrial Biological Research Association*, *47*(8), 2097–2102. https://doi.org/10.1016/j.fct.2009.05.031

Kumar, D., Geethanjali, B., Ko, A., Kumar, Gk, R., & Basalingappa, K. M. (2017). Tinospora cordifolia: The Antimicrobial Property of the Leaves of Amruthaballi. *Journal of Bacteriology & Mycology: Open Access*, *5*(5),363-371. DOI: 10.15406/jbmoa.2017.05.00147

Kumar, V., Singh, S., Singh, A., Dixit, A. K., Srivastava, B., Sidhu, G. K., Singh, R. P., Meena, A. K., Singh, R. K., Subhose, V., & Prakash, O. (2018). Phytochemical, Antioxidant, Antimicrobial, and Protein Binding Qualities of Hydro-ethanolic Extract of *Tinospora cordifolia*. *Journal of Biologically Active Products from Nature, 8*(3), 192-200, DOI: 10.1080/22311866.2018.1485513.

Li, Y., Zhang, T., Huai, J., Cheng, C., Xie, L., Wang, S., & Dai, R. (2021). Pharmacokinetic study of three different formulations of l-tetrahydropalmatine in brain tissues of rats. *Biomedical chromatography: BMC*, *35*(5), e5066. https://doi.org/10.1002/bmc.5066.

Li, Z., Geng, Y. N., Jiang, J. D., & Kong, W. J. (2014). Antioxidant and anti-inflam-matory activities of berberine in the treatment of diabetes mellitus. *Evidence-based complementary and alternative medicine: eCAM*, *2014*, 289264. https://doi.org/10.1155/2014/289264.

Lin, M. T., Chueh, F. Y., & Hsieh, M. T. (2001). The hypothermic effects of dl-tetrahydropalmatine in rats. *Neuroscience letters*, *315*(1-2), 53–56. https://doi.org/10.1016/s0304-3940(01)02323-0.

Lin, Z., Li, S., Guo, P., Wang, L., Zheng, L., Yan, Z., Chen, X., Cheng, Z., Yan, H., Zheng, C., & Zhao, C. (2019). Columbamine suppresses hepatocellular carcinoma cells through down-regulation of PI3K/AKT, p38 and ERK1/2 MAPK signaling pathways. *Life sciences*, *218*, 197–204. https://doi.org/10.1016/j.lfs.2018.12.038.

Liu, D., Meng, X., Wu, D., Qiu, Z., & Luo, H. (2019). A Natural Isoquinoline Alkaloid with Antitumor Activity: Studies of the Biological Activities of Berberine. *Frontiers in pharmacology*, *10*, 9. https://doi.org/10.3389/fphar.2019.00009.

Liu, J., Li, Q., Wang, C., Shao, J., Wang, T., Wu, D., Ma, K., Yan, G., & Yin, D. (2020). Antifungal evaluation of traditional herbal monomers and their potential for inducing cell wall remodeling in *Candida albicans* and *Candida auris*. *Biofouling*, *36*(3), 319–331. https://doi.org/10.1080/08927014.2020.1759559.

Liu, X., Zhang, X., Ye, L., & Yuan, H. (2016). Protective mechanisms of berberine against experimental autoimmune myocarditis in a rat model. *Biomedicine & pharmacotherapy = Biomedecine & pharmacotherapie*, *79*, 222–230. https://doi.org/10.1016/j.biopha.2016.02.015.

Long, J., Song, J., Zhong, L., Liao, Y., Liu, L., & Li, X. (2019). Palmatine: A review of its pharmacology, toxicity and pharmacokinetics. *Biochimie*, *162*, 176–184. https://doi.org/10.1016/j.biochi.2019.04.008.

Luo, N., Jin, L., Yang, C., Zhu, Y., Ye, X., Li, X., & Zhang, B. (2021). Antifungal activity and potential mechanism of magnoflorine against *Trichophyton rubrum*. *The Journal of antibiotics*, *74*(3), 206–214. https://doi.org/10.1038/s41429-020-00380-4.

Meena, A. K., Singh, A., Panda, P., Mishra, S., Rao, M. M. (2010). *Tinospora cordifolia*: its bioactivities & evaluation of physicochemical properties. *International Journal of Pharmacognosy and Phytochemical Research*, 2(2),50-55. https://ijppr.com/volume2issue2/.

Modi, B., Shah, K. K., Shrestha, J., Shrestha, P., Basnet, A., Tiwari, I., & Aryal, S. P. (2021). Morphology, Biological Activity, Chemical Composition, and Medicinal Value of *Tinospora cordifolia* (willd.) Miers. *Advanced Journal of Chemistry, Section B: Natural Products and Medical Chemistry*.*3*(1), 36-54. DOI: 10.22034/AJCB.2021.118153.

Neag, M. A., Mocan, A., Echeverría, J., Pop, R. M., Bocsan, C. I., Crişan, G., & Buzoianu, A. D. (2018). Berberine: Botanical Occurrence, Traditional Uses, Extraction Methods, and Relevance in Cardiovascular, Metabolic, Hepatic, and Renal Disorders. *Frontiers in pharmacology*, *9*, 557. https://doi.org/10.3389/fphar.2018.00557.

Newman, D. J., & Cragg, G. M. (2012). Natural products as sources of new drugs over the 30 years from 1981 to 2010. *Journal of natural products*, *75*(3), 311–335. https://doi.org/10.1021/np200906s.

Och, A., Podgórski, R., & Nowak, R. (2020). Biological Activity of Berberine-A Summary Update. *Toxins*, *12*(11), 713. https://doi.org/10.3390/toxins12110713.

Okon, E., Kukula-Koch, W., Halasa, M., Jarzab, A., Baran, M., Dmoszynska-Graniczka, M., Angelis, A., Kalpoutzakis, E., Guz, M., Stepulak, A., & Wawruszak, A. (2020). Magnoflorine-Isolation and the Anticancer Potential against NCI-H1299 Lung, MDA-MB-468 Breast, T98G Glioma, and TE671 Rhabdomyosarcoma Cancer Cells. *Biomolecules*, *10*(11), 1532. https://doi.org/10.3390/biom10111532.

Okoń, E., Luszczki, J. J., Kukula-Koch, W., Halasa, M., Jarzab, A., Khurelbat, D., Stepulak, A., & Wawruszak, A. (2020). Synergistic or Additive Pharmacological Interactions between Magnoflorine and Cisplatin in Human Cancer Cells of Different Histological Origin. *International journal of molecular sciences*, *21*(8), 2848. https://doi.org/10.3390/ijms21082848.

Patel, M. B., & Mishra, S. (2012). Isoquinoline alkaloids from *Tinospora cordifolia* inhibit rat lens aldose reductase. *Phytotherapy research: PTR*, *26*(9), 1342–1347. https://doi.org/10.1002/ptr.3721.

Phan, V. K., Chau, V. M., Nguyen, T. D., La, V. K., Dan, T. H., Nguyen, H. N., Nguyen, X. C., Hoang, T. H., & Trinh, V. L. (2010). Aporphine alkaloids, clerodane diterpenes, and other constituents from Tinospora cordifolia. *Fitoterapia, 81*(6), 485–489. https://doi.org/10.1016/j.fitote.2010.01.005.

Rolle, J., Asante, D. O., Kok-Fong, L. L., Boucetta, H., Seidu, T. A., Tai, L., & Alolga, R. N. (2021). Jatrorrhizine: a review of its pharmacological effects. *The Journal of pharmacy and pharmacology*, *73*(6), 709–719. https://doi.org/10.1093/jpp/rgaa065.

Semwal, D. K., Badoni, R., Semwal, R., Kothiyal, S. K., Singh, G. J., & Rawat, U. (2010). The genus *Stephania* (Menispermaceae): chemical and pharmacological perspectives. *Journal of ethnopharmacology*, *132*(2), 369–383. https://doi.org/10.1016/j.jep.2010.08.047.

Sharma, P., Dwivedee, B. P., Bisht, D., Dash, A. K., & Kumar, D. (2019). The chemical constituents and diverse pharmacological importance of *Tinospora cordifolia*. *Heliyon*, *5*(9), e02437. https://doi.org/10.1016/j.heliyon.2019.e02437.

Sharma, U., Bala, M., Kumar, N., Singh, B., Munshi, R. K., & Bhalerao, S. (2012). Immunomodulatory active compounds from *Tinospora cordifolia*. *Journal of ethnopharmacology*, *141*(3), 918–926. https://doi.org/10.1016/j.jep.2012.03.027.

Shen, Y., Guan, S., Ge, H., Xiong, W., He, L., Liu, L., Yin, C., Liu, H., Li, G., Xu, C., Xu, H., Liu, S., Li, G., Liang, S., & Gao, Y. (2018). Effects of palmatine on rats with comorbidity of diabetic neuropathic pain and depression. *Brain research bulletin*, *139*, 56–66. https://doi.org/10.1016/j.brainresbull.2018.02.005.

Singh, D., & Chaudhuri, P. K. (2017). Chemistry and Pharmacology of *Tinospora cordifolia*. *Natural Product Communications*. *12*(2), 299–308. https://doi.org/10.1177/1934578X1701200240.

Stegelmeier, B. L., Edgar, J. A., Colegate, S. M., Gardner, D. R., Schoch, T. K., Coulombe, R. A., & Molyneux, R. J. (1999). Pyrrolizidine alkaloid plants, metabolism and toxicity. *Journal of natural toxins*, *8*(1), 95–116.

Sun, X. L., Zhang, X. W., Zhai, H. J., Zhang, D., & Ma, S. Y. (2020). Magnoflorine inhibits human gastric cancer progression by inducing autophagy, apoptosis and cell cycle arrest by JNK activation regulated by ROS. *Biomedicine & pharmacotherapy, 125*, 109118. https://doi.org/10.1016/j.biopha.2019.109118.

Tarabasz, D., & Kukula-Koch, W. (2020). Palmatine: A review of pharmacological properties and pharmacokinetics. *Phytotherapy research: PTR, 34*(1), 33–50. https://doi.org/10.1002/ptr.6504.

Tsai, S. F., & Lee, S. S. (2010). Characterization of Acetylcholinesterase Inhibitory Constituents from *Annona glabra* Assisted by HPLC Microfractionation. *Journal of natural products, 73*(10), 1632–1635. https://doi.org/10.1021/np100247r.

Upadhyay, A. K., Kumar, K., Kumar, A., & Mishra, H. S. (2010). *Tinospora cordifolia* (Willd.) Hook, F. and Thoms. (*Guduchi*) - validation of the Ayurvedic pharmacology through experimental and clinical studies. *International journal of Ayurveda research, 1*(2), 112–121. https://doi.org/10.4103/0974-7788.64405.

Veeresham C. (2012). Natural products derived from plants as a source of drugs. *Journal of advanced pharmaceutical technology & research, 3*(4), 200–201. https://doi.org/10.4103/2231-4040.104709.

Wang, L. J., Jiang, Z. M., Xiao, P. T., Sun, J. B., Bi, Z. M., & Liu, E. H. (2019). Identification of anti-inflammatory components in Sinomenii caulis based on spectrum-effect relationship and chemometric methods. *Journal of pharmaceutical and biomedical analysis, 167*, 38–48. https://doi.org/10.1016/j.jpba.2019.01.047.

Wang, N., Tan, H. Y., Li, L., Yuen, M. F., & Feng, Y. (2015). Berberine and Coptidis Rhizoma as potential anticancer agents: Recent updates and future perspectives. *Journal of ethnopharmacology, 176*, 35–48. https://doi.org/10.1016/j.jep.2015.10.028.

Wang, Y., Liu, Y., Du, X., Ma, H., & Yao, J. (2020). The Anti-Cancer Mechanisms of Berberine: A Review. *Cancer management and research, 12*, 695–702. https://doi.org/10.2147/CMAR.S242329.

Wei, T., Xiaojun, X., & Peilong, C. (2020). Magnoflorine improves sensitivity to doxorubicin (DOX) of breast cancer cells via inducing apoptosis and autophagy through AKT/mTOR and p38 signaling pathways. *Biomedicine & pharmacotherapy = Biomedecine & pharmacotherapie, 121*, 109139. https://doi.org/10.1016/j.biopha.2019.109139.

Yu, H., Wang, Y., Wang, X., Guo, J., Wang, H., Zhang, H., & Du, F. (2019). Jatror-rhizine suppresses the antimicrobial resistance of methicillin-resistant *Staphylococcus aureus*. *Experimental and therapeutic medicine, 18*(5), 3715–3722. https://doi.org/10.3892/etm.2019.8034.

Zhang, X. X., Tian, Y., Wang, Z. T., Ma, Y. H., Tan, L., & Yu, J. T. (2021). The Epidemiology of Alzheimer's Disease Modifiable Risk Factors and Prevention. *The journal of prevention of Alzheimer's disease, 8*(3), 313–321. https://doi.org/10.14283/jpad.2021.15.

Zhong, F., Chen, Y., Chen, J., Liao, H., Li, Y., & Ma, Y. (2022). Jatrorrhizine: A Review of Sources, Pharmacology, Pharmacokinetics and Toxicity. *Frontiers in pharmacology*, *12*, 783127. https://doi.org/10.3389/fphar.2021.783127.

Zhong, M., Liu, Y., Liu, J., Di, D., Xu, M., Yang, Y., Li, W., Chen, Y., & Liu, J. (2014). Isocorydine derivatives and their anticancer activities. *Molecules (Basel, Switzerland)*, *19*(8), 12099–12115. https://doi.org/10.3390/molecules190812099.

Chapter 8

Theophylline: A Bioactive Dimethylxanthine Alkaloid

Santwana Palai[1], Subhash Chandra[2,3,*] Neetu Pandey [2] and Rajbeer Singh[2]

[1]Department of Veterinary Pharmacology & Toxicology,
College of Veterinary Science and Animal Husbandry,
Odisha University of Agriculture and Technology, Odisha, India
[2]Department of Applied Chemistry and Basic Science, Sardar Bhagwan Singh University Balawala Dehradun, Uttarakhand, India
[3]Department of Pharmaceutical Chemistry, School of Sciences,
Hemvati Nandan Bahuguna Garhwal University (A Central University),
Uttarakhand, India

Abstract

Theophylline "1,3-dimethylxanthine" is present in trace amounts in tea leaves, coffee, green coffee, unsweetened natural cocoa, and mate. It acts as an anti-inflammatory, antioxidant, and immunomodulator. Theophylline was found to have anti-inflammatory effects on cytokines produced primarily by mononuclear phagocytic cells. It increases steroid sensitivity and thus decreases oxidative stress. It dilates the airways of the lungs for ease of breathing. It lowers the bronchial airways and smooths muscle activity and blood vessels in the lungs. It also reduces the airway's sensitivity to adenosine, histamine, methacholine, and allergens. It's been used for a long time for the treatment of respiratory diseases, singly or combinedly with other medications. Shortness of breath, chest tightness, and wheezing are triggered in chronic bronchitis,

* Corresponding Author's Email: subhashkothiyal@gmail.com.

In: The Essential Guide to Alkaloids
Editor: Deepak Kumar Semwal
ISBN: 979-8-88697-456-0

and emphysema, asthma-like lung disease is cured through theophylline. The swift rise of the COVID-19 pandemic necessitated the advancement in treatments for rightly attacking new variants of the SARS-CoV-2 virus. Theophylline can be a pertinent option as it targets numerous immunological targets of COVID-19 pathophysiology. This chapter sorts effects and applications of theophylline having potential bronchodilator, immunomodulatory and antiviral effects which makes it an interesting candidate in the treatment of COVID-19.

Keywords: antioxidant, bronchodilatation, inflammation, immunomodulation methylxanthine, phosphodiesterase

1. Introduction

Theophylline "1,3-dimethyl-7H-purine-2,6-dione" is a dimethylxanthine with two methyl groups derived from the xanthine purine base at positions 1 and 3. It is naturally found in black tea, coffee, chocolate, dried mate, and other foods. Shortness of breath, wheezing-like symptoms of chronic bronchitis, and emphysema are treated with theophylline. It is a phosphodiesterase inhibitor which can successfully treat respiratory diseases like chronic obstructive pulmonary diseases (Hondo et al., 2022).

Theophylline relaxes smooth muscle and causes noteworthy bronchodilation. It possesses a positive ionotropic effect, cardiac stimulant, CNS stimulant and mild diuretic properties. Theophylline act as a phosphodiesterase inhibitor and adenosine antagonist and thus is primarily responsible for its relaxation of bronchial smooth muscle and pulmonary blood vessels. It is a real anti-inflammatory and immunomodulatory agent. Because inexpensive and widely available, theophylline is most commonly prescribed in the therapy of asthma and COPD worldwide. Theophylline is active in the long-term management of respiratory diseases with inflammation like asthma and COPD (Banner and Page, 1995).

Theophylline has many antiviral properties that could impede virus pathogenesis. Many viral respiratory infections caused by MERS, influenza, and SARS-CoV viruses elicit similar host immune responses. So, theophylline can be a candidate in the drug discovery process against viral pathogens causing several viral diseases. Theophylline inhibits the activation of the inflammatory processes induced by SARS-CoV-2 via a variety of mechanisms. So, theophylline can be a promising candidate in the treatment of COVID-19 because it targets some immunological aims tangled in the disease's physiology and pathology. Theophylline is effective in treating sinus bradycardia, which is the most common arrhythmia found in COVID-19 caused as a side effect of COVID-19 (Dahiya et al., 2022).

After the novel mechanisms of action accounting for theophylline's effectiveness in severe asthma was identified, attempts have been made to minimise the side effects of theophylline. This selective phosphodiesterase (PDE) inhibitor has the potential to exert anti-inflammatory, anti-oxidative and bronchodilatory effects to improve the beneficial effect in lung diseases. The anti-inflammatory properties of the phosphodiesterase inhibitors can inhibit the activation of inflammatory cells like T-lymphocyte, eosinophils, macrophages, mast cells etc. in COVID-19 pneumonia too (Wall et al., 2021).

2. History, Chemistry, Source and Therapeutic Uses

2.1. History

Theophylline was recovered from tea leaves in 1888 and chemically recognized by the German biologist Albrecht Kossel. After 7 years, Emil Fischer and Lorenz Ach described a chemical amalgamation on basis of 1,3-dimethyluric acid. Wilhelm Traube, another German scientist, invented the Traube purine synthesis in 1900 as a different method of synthesising theophylline. Theophylline was first clinically used as a diuretic in 1902. In 1922, it was conveyed as an effective treatment for asthma (Wainwright et al., 2022).

2.2. Chemistry

Theophylline (Figure 1) being a methylxanthine is structurally alike to dietary xanthines like caffeine, theobromine etc. Although several substituted derivatives are synthesised, nothing has a better performance than theophylline. Enprofylline is a 3-propyl derivative having added potency as a bronchodilator, but then again, its clinical progress was stopped due to liver toxicity. Many theophylline salts are commercialised, such as aminophylline, an ethylene diamine salt that has increased solubility at neutral pH, allowing it to be administered intravenously. Doxofylline is (7-(1,3-dioxalan-2-yl-methyl) theophylline) analogous to theophylline in effectiveness with fewer effects on adenosine receptors and thus lesser central and cardiovascular adverse properties. Choline theophyllinate possesses no benefit and acepifylline is inactive (Kunisaki and Sin, 2021).

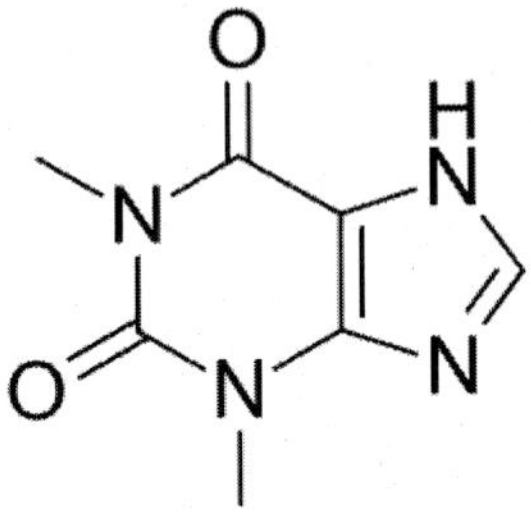

Figure 1. Chemical structure of Theophylline.

2.3. Source

Table 1. Natural sources of theophylline alkaloid

S.No	Scientific name and family	Common name and part used	Geographical location	Therapeutic and other uses	Ref.
1	*Theobroma cacao* L (Malvaceae)	Cocoa beans (seeds)	Central America, Caribbean islands, South American states	1. Chocolate and chocolate confections 2. Diarrhoea, asthma and bronchitis	Sioriki et al., 2022
2	*Camellia sinensis* (L) Kuntze (Theaceae)	Green tea (leaves and leaf buds)	East Asia, China, Burma	1. Delicate flavor drink 2. Anti-inflammatory and immunomodulatory	Sonia and Malik, 2022
3	*Paullinia cupana* HB Kunth (Sapindaceae)	Brazilian cocoa, zoom (crushed seeds)	Amazon basin, Brazil	1. Energy drinks stimulant 2. Weight loss, pain relief and healthier skin	Santana and Macedo, 2019
4	*Cola nitida* (Vent) Schott & Endl (Sterculiaceae)	Kola, bitter kola (nut)	Tropical Africa, America	1. Herbal supplement 2. Bronchodilator, CNS stimulant and antidiarrheal	Olalere et al., 2021
5	*Cola acuminata* P. Beauv (Sterculiaceae)	Kola nut (nut)	Tropical Africa, America	1. Flavouring agent 2. Fatigue and depression	Bamigbola et al., 2018

Figure 2. Selected plant species as sources of theophylline alkaloid.

Theophylline is naturally created in tea and cocoa. This xanthine family member is a byproduct of caffeine metabolic processing in the liver. Cocoa beans contain theophylline naturally. Criollo cocoa beans have been found to contain as much as 3.7 mg/g. Although brewed tea contains trace amounts of theophylline, it only provides about 1 mg/l, which is lesser than the beneficial dosage. Theophylline is also present in trace amounts in *Paullinia cupana* (guarana) and *Cola acuminata* and *Cola nitida* (kola nuts/cola plant) (Table 1; Figure 2) (Ali et al., 2022; Ali, 2022).

2.4. Therapeutic Uses of Theophylline

Theophylline (1,3-dimethylxanthine) is primarily used to treat asthma and chronic obstructive pulmonary disease as a bronchodilator. Premature newborns are commonly treated with theophylline for bradycardia and apnea. Theophylline induces endogenous catecholamine release via indirect activation of β-1 & 2 receptors leading to bronchodilation. Acebrophylline, an anti-inflammatory and airway mucoregulator, is a combination of ambroxol and theophylline (Monteiro et al., 2014).

3. Pharmacodynamics and Pharmacokinetics

3.1. Pharmacodynamics

Despite significant recent advances, the cellular mechanism of action and site of action of theophylline remain unknown after more than 70 years of clinical use. The proposed molecular mechanisms of action occur significantly at higher effective theophylline concentrations. (Figure 3) (Montaño et al., 2022).

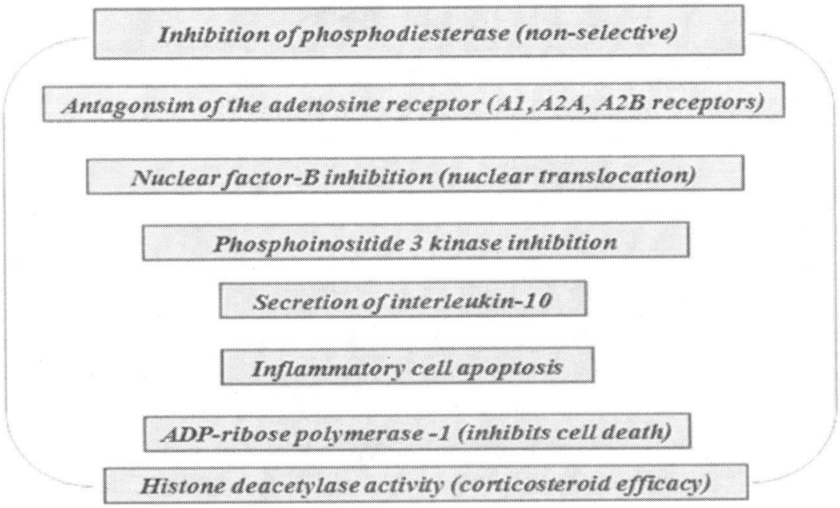

Figure 3. Proposed mechanisms of action of theophylline.

3.2. Pharmacokinetics

Following oral intake, theophylline is swiftly and entirely captivated in the gastrointestinal tract. This alkaloid is extensively metabolised by at least two cytochrome P450 isozymes in the liver and has no significant pre-systemic elimination. Theophylline has a 7–9-hour serum half-life and is toxic at doses of 7.5 mg/kg and higher. When administering theophylline in cases of reduced drug clearance like heart failure, and hepatic or pulmonary ailments (pneumonia, COPD) special precautions must be taken. This alkaloid has no significant pre-systemic elimination and freely allocates into fat-free tissues. It undergoes expansive metabolism through at least two cytochrome P450 isozymes in the liver. The serum half-life of theophylline varies approximately from 3 to 13, the average being 7-9 hours but the doses of 7.5 mg/kg or higher can be toxic. Children under the age of 16 and smokers should expect higher clearance (Verbesselt et al., 1979).

4. Drug Interaction

Many drug interactions should be considered when administering theophylline, due to decreased clearance when combined with 5-lipoxygenase inhibitor zileuton, quinolones, allopurinol, erythromycin, cimetidine, and serotonin uptake inhibitors, and increased metabolism when combined with rifampicin, phenytoin and phenobarbital (Barnes, 2013).

5. Spectroscopic Characterisation

It is soluble in 0.1N NaOH and has an extinction coefficient of 10,200 (cm1 M1) at 277 nm in UV. The typical signals that distinguish theophylline from likely methylxanthines are around δ3.23 and 3.41 in ^{1}H-NMR, due to theophylline's sole methylation. At δ 8.01, the enduring proton signal corresponds to the proton on the imidazole ring that was not transferred between the nitrogen. The transferred proton between the nitrogen atoms is a variable proton that exhibits a signal sometimes. Theophylline's unique methylation agrees with δ 27.7 and 29.9 signals in the ^{13}C-NMR. The remaining signals resemble the carbons found in the xanthine support (Suryana et al., 2019).

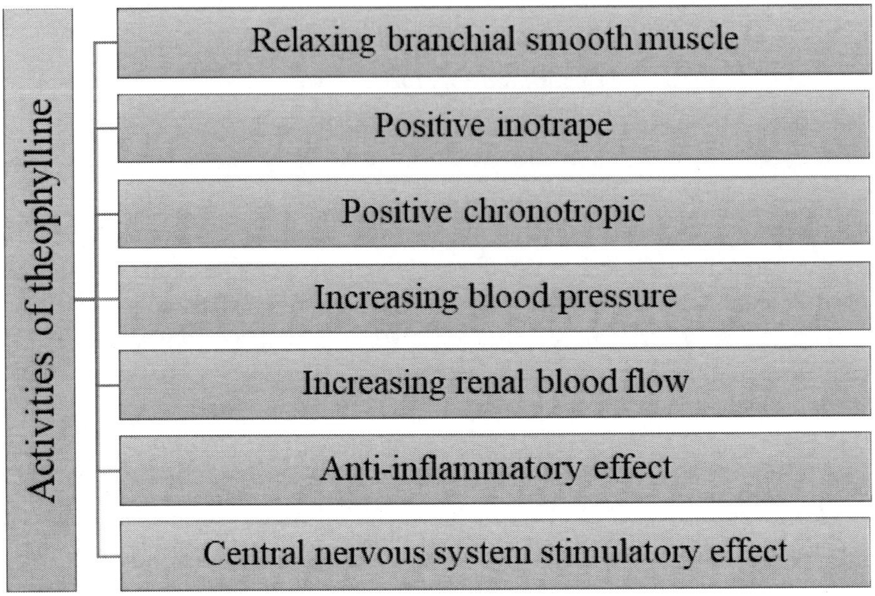

Figure 4. Different actions of theophylline.

6. Effect of Theophylline on Various Systems

Theophylline has bronchodilatory, immunomodulatory, anti-inflammatory, and antiviral properties. The main actions of theophylline are given below.

6.1. Effect on Bronchial System

Along with its immunomodulatory actions, theophylline acts as a bronchodilator that lowers mean pulmonary arterial pressure *in vivo*. This drug's pharmacodynamics is still being debated. Phosphodiesterase (PDE) inhibition and adenosine antagonism, like other actions, have been proposed to clarify its efficiency in the treatment of asthma. Multiple PDE isoenzymes catalyse the breakdown of cAMP and cGMP, which are involved in smooth muscle tone regulation. PDE isoenzymes have been identified in the human bronchus and pulmonary artery, and the pharmacological actions of PDE inhibitors were explored. Theophylline and selective PDE III inhibitors relax human bronchus and pulmonary arteries, while PDE IV inhibitors relax precontracted bronchus and PDE V/I inhibitors help to relax the pulmonary artery. Synergy exists between PDE III and PDE IV inhibitors for relaxing the bronchus, and between PDE III and PDE V inhibitors in relaxing the pulmonary artery.

The close relationship between theophylline concentrations preventing bronchus or pulmonary artery PDE and those producing relaxation suggests that PDE inhibition is an important mechanism of action of theophylline in smooth muscle relaxation (Rabe et al., 1995).

6.2. Effect on COVID-19

Theophylline and dyphylline are methylxanthines used to treat airway obstruction in asthma, emphysema, infant apnea, chronic bronchitis and chronic obstructive pulmonary disease and are repurposed for use in the treatment of COVID-19. Wall et al., conducted retrospective research on COVID-19 patients who needed oxygen. An increased ROX score (ratio of oxygen saturation measured by pulse oximetry/FiO2 to respiratory rate) and decreased C-reactive protein and mortality in the theophylline/pentoxifylline group indicated that this treatment may benefit COVID-19 patients. Theophylline targets several immunological targets involved in disease pathophysiology. Theophylline inhibits more activation of the inflammatory progressions induced by SARS-CoV-2 via a variety of mechanisms. Furthermore, it may have antiviral properties that could prevent the pathogenesis of influenza, SARS-CoV, MERS, or SARS-CoV-2 viruses. Theophylline, which has a long history of clinical use and is inexpensive, could be a viable treatment option for COVID-19 patients (Montaño et al., 2022).

6.3. Effect on the Cardiovascular System

Theophylline is an older drug that has bronchial dilatation, diuresis, and chronotropic and dromotropic effects. Theophylline was effective in treating sick sinus syndrome, increasing heart frequency and avoiding permanent pacemaker implantation. Thus, oral theophylline is a viable alternative to permanent pacemaker implantations in patients with increased heart rates following heart transplantation, allowing for the withdrawal of chronotropic support and avoiding the need for permanent pacemaker implantation (Kooij et al., 2022). This sinus bradycardia (common arrhythmias) found in COVID-19-affected persons may be caused by viral myocarditis, myocardial ischemia, or as an adverse effect of COVID-19 medicine (Dahiya et al., 2022).

6.4. Effect on Renal System

Theophylline relaxes smooth muscle and causes noteworthy bronchodilation, has a positive ionotropic action, cardiac and CNS stimulant and mild diuretic properties. SARSCoV2 fixes to ACE2 receptors in various renal organs, particularly kidney proximal tubules with more ACE2 receptors than the lungs affecting renal tubules. Theophylline can help with the massive albuminuria and subsequent proteinuria seen in SARS-CoV-2 (Ahmadian et al., 2021).

6.5. Effect on Adipose Tissue

Lipase inhibitors, an effective anti-obesity drug, act on the intestinal tract with no central nervous system side effects. The lipase inhibitor Fu Brick Theophylline (FBT) was partitioned using enzyme molecular dynamics, and lipase inhibitors' inhibition mechanism on obesity was investigated. It inhibited lipase with an IC50 of 1.020.03 g/mL in 3T3-L1 preadipocytes, flow cytometry to sense apoptosis, and intracellular ROS action in preadipocytes. During cell differentiation, Fu Brick Theophylline could decrease ROS creation and inflammation factor mRNA expression. The morphological observation and gene expression analysis of high-fat feed nutritional obese mice model in terms of adipocyte and hepatocyte metabolism and body weight mice demonstrated FBT can efficiently decrease fat accumulation, obesity, degree of fatty liver, liver fibrosis, and advance mitochondrial membrane structure injury (Liu et al., 2022).

7. Adverse Effects/Toxicity

Theophylline causes complications due to its narrow therapeutic window. Hyperglycemia, arrhythmias, rhabdomyolysis, and seizures are some adverse effects that can occur in patients with theophylline toxicity (Sharma et al., 2022). It can also cause nausea and diarrhoea, as well as abnormal heart rhythms, increased heart rate and CNS excitation (light-headedness, insomnia, irritability, dizziness and headaches). As a result of a neurological emergency, seizures and tachyarrhythmias can occur. Erythromycin, cimetidine, fluoroquinolones like ciprofloxacin, and SSRIs like fluvoxamine increase its toxicity. Beta-blockers can be used to treat theophylline toxicity (Zhou et al., 2006). In theophylline toxicity, epinephrine levels increase and increased catecholamine concentrations cause metabolic acidosis, hyperglycemia and hypokalemia cardiac arrhythmias (Greene et al., 2018).

8. Uses under Investigation

Clinical research in 2008 found that theophylline improved the sense of smell in anosmia patients (Jahandideh et al., 2022). Also, another clinical trial in 2004 compared theophylline to methylphenidate in the treatment of attention deficit hyperactivity disorder in children and adolescents (Childress and Burton, 2022). The in-silico structure-based techniques were used to determine whether theophylline is a viable candidate for use in the treatment of COVID-19. Several revisions have looked into the interactions of theophylline with various proteins. The chymotrypsin-like protease protein (3CLpro) and the nucleocapsid are two examples of target proteins. The collaboration between the spike protein and ACE2 makes them potential pharmacologic goals. Another methylxanthine caffeine is also being investigated as a potential pharmacological means (Montaño et al., 2022).

Conclusion

Theophylline was previously used to treat airway obstruction in lung disorders such as emphysema, asthma, infant apnea, COPD and chronic bronchitis due to its bronchodilatory, immunomodulatory, anti-inflammatory, and antiviral properties. Theophylline has recently been repurposed to treat coronavirus disease 2019. Structure-based techniques also showed its efficacy against SARS-CoV-2. Thus, theophylline can be used alone or in combination with existing drugs to treat COVID-19 symptoms.

References

Ahmadian E., Hosseiniyan Khatibi S. M., Razi Soofiyani S., Abediazar S., Shoja M. M., Ardalan M., Zununi Vahed S. COVID-19 and kidney injury: Pathophysiology and Molecular Mechanisms. *Reviews in Medical Virology,* 2021, 31:e2176.

Ali A., Zahid H. F., Cottrell J. J., Dunshea F. R. A Comparative Study for Nutritional and Phytochemical Profiling of Coffea arabica (C. arabica) from Different Origins and Their Antioxidant Potential and Molecular Docking. *Molecules,* 2022, 27(16):5126.

Ali R. Green Tea Catechins: Functionality, Addition to Food, and Bioavailability. In *Tea as a Food Ingredient*, CRC Press, 2022, pp. 33-69.

Bamigbola E. A., Attama A. A., Ogeh P. C. Evaluation of physico-mechanical and mucoadhesive properties of biopolymer films from Cola acuminata gum. *Nigerian Journal of Pharmaceutical Research*, 2018, 14(1):1-3.

Banner K. H., Page C. P. Theophylline and selective phosphodiesterase inhibitors as anti-inflammatory drugs in the treatment of bronchial asthma. *European Respiratory Journal*, 1995, 8(6):996-1000.

Barnes P. J. Theophylline. American Journal of Respiratory and Critical Care Medicine, 2013, 188:901-906.

Childress A., Burton S. Evaluating the pharmacokinetics of extended release viloxazine in the treatment of children with attention-deficit/hyperactivity disorder. *Expert Opinion on Drug Metabolism & Toxicology*, 2022, 18(6):357-366.

Dahiya A., Sharma R., Singh A., Joshi P., Wardhan H. Role of Etophylline and Theophylline Prolonged Release Tablet in COVID-19 Associated Sinus Node Dysfunction. *Journal of the Association of Physicians of India,* 2022, 70, 11-12.

Greene S. C., Halmer T., Carey J. M., Rissmiller B. J., Musick M. A. Theophylline toxicity: An old poisoning for a new generation of physicians. *Turkish Journal of Emergency Medicine*, 2018, 18(1):37-39.

Hondo H., Kowase S., Asano S., Osada J., Aoki H., Yumoto K. Theophylline to treat prolonged paroxysmal complete atrioventricular block without conduction disorder or structural heart disease after COVID-19 infection: A case report. *HeartRhythm Case Reports*, 2022, 8(4):229-32.

Jahandideh H., Rahimi P., Mobarakeh P. F., Firouzabadi F. D., Roomiani M., Farhadi M. The Effect of Topical Aminophylline on Hyposmia and Anosmia. *Iranian Journal of Otorhinolaryngology*, 2022, 34(123):171.

Kooij C., Szili-Torok T., Roest S., Constantinescu A. A., Brugts J. J., Manintveld O., Caliskan K. Theophylline Use to Prevent Permanent Pacing in the Contemporary Era of Heart Transplantation: The Rotterdam Experience. *Frontiers in Cardiovascular Medicine*, 2022, 9, 896141.

Kunisaki K. M., Sin D. D. Methylxanthines in COPD: yes to caffeine, no to theophylline. *European Respiratory Journal,* 2021, 57(6), 2004564.

Liu T. T., Liu X. T., Huang G. L., Liu L., Chen Q. X., Wang Q. Theophylline Extracted from Fu Brick Tea Affects the Metabolism of Preadipocytes and Body Fat in Mice as a Pancreatic Lipase Inhibitor. *International Journal of Molecular Sciences*, 2022, 23(5):2525.

Montaño L. M., Sommer B., Gomez-Verjan J. C., Morales-Paoli G. S., Ramírez-Salinas G. L., Solís-Chagoyán H., Sanchez-Florentino Z. A., Calixto E., Pérez-Figueroa G. E., Carter R., Jaimez-Melgoza R. Theophylline: Old Drug in a New Light, Application in COVID-19 through Computational Studies. *International Journal of Molecular Sciences*. 2022, 23(8):4167.

Monteiro J., Alves M. G., Oliveira P. F., Silva B. M. Pharmacological potential of methylxanthines: Retrospective analysis and future expectations. *Critical Reviews in Food Science and Nutrition*, 2019, 59(16):2597-625.

Olalere O. A., Gan C. Y., Akintomiwa O. E., Adeyi O., Adeyi A. Optimisation of microwave-assisted extraction and functional elucidation of bioactive compounds from Cola nitida pod. *Phytochemical Analysis,* 2021, 32(5):850-8.

Rabe K. F., Magnussen H., Dent G. Theophylline and selective PDE inhibitors as bronchodilators and smooth muscle relaxants. *European Respiratory Journal*, 1995, 8(4):637-42.

Santana Á. L., Macedo G. A. Effects of hydroalcoholic and enzyme-assisted extraction processes on the recovery of catechins and methylxanthines from crude and waste seeds of guarana (Paullinia cupana). *Food chemistry,* 2019, 281:222-30.

Sharma S., Hashmi M. F., Chakraborty R. K. Asthma Medications. 2022 Jun 23. In: *StatPearls Treasure Island* (FL): StatPearls Publishing; 2022. PMID: 30285350.

Sioriki E., Tuenter E., Van de Walle D., Lemarcq V., Cazin C. S., Nolan S. P., Pieters L., Dewettinck K. The effect of cocoa alkalization on the non-volatile and volatile mood-enhancing compounds. *Food Chemistry*, 2022, 381:132082.

Sonia R., Malik S. Therapeutic Properties of Green Tea: A Review. *Journal of Multidisciplinary Applied Natural Science,* 2022, 2(2):90-102.

Suryana S., Nurjanah N. S., Permana B., Prasetiawati R., Lubis N. Derivative Uv-Spectroscopic determination of theophylline, salbutamol sulfate and glycerylguaicolate in syrup mixture. *Journal of Physics: Conference Series,* 2019, 1402(5):055043).

Verbesselt R., Tjandramaga T. B., De Schepper P. J. Simultaneous determination of theophylline and its metabolites in urine. *Current Medical Research and Opinion*, 1979, 6(sup6):151-60.

Wainwright C. L., Teixeira M. M., Adelson D. L., Buenz E. J., David B., Glaser K. B., Harata-Lee Y., Howes M. J., Izzo A. A., Maffia P., Mayer A. M. Future directions for the discovery of natural product-derived immunomodulating drugs. *Pharmacological research*, 2022, 21:106076.

Wall G. C., Smith H. L., Trump M. W., Mohr J. D., DuMontier S. P., Sabates B. L., Ganapathiraju I., Kable T. J. Pentoxifylline or theophylline use in hospitalized COVID-19 patients requiring oxygen support. *Clinical Respiratory Journal,* 2021, 15(7):843-846.

Zhou Y., Wang X., Zeng X., Qiu R., Xie J., Liu S., Zheng J., Zhong N., Ran P. Positive benefits of theophylline in a randomized, double-blind, parallel-group, placebo-controlled study of low-dose, slow-release theophylline in the treatment of COPD for 1 year. *Respirology,* 2006, 11:603-610.

Chapter 9

Alkaloids as Botanical Pesticides for Plants Protection

**Himani Karakoti, Sonu Kumar Mahawer, Tanuja Kabdal, Ravendra Kumar*
and Om Prakash**

Department of Chemistry, College of Basic Sciences and Humanities,
G.B. Pant University of Agriculture and Technology, Pantnagar,
U.S. Nagar, Uttarakhand, India

Abstract

Alkaloids are the most basic, naturally occurring nitrogen-containing organic compounds frequently found in plants, animals and microbes. These are the important secondary metabolites known to have a number of therapeutic and pharmacological properties. Several studies have focused on the wide range of pharmaceutical potential of alkaloids from various plants. In recent years, research regarding the pesticidal properties of alkaloids has gradually increased as they are safer for the environment, easily degradable and have low toxicity as compared to chemical pesticides. This book chapter is mainly focusing on the different properties of alkaloids as a botanical pesticide to prevent insect pests and diseases.

Keywords: alkaloids, insecticide, antimicrobial, nematicidal

---

* Corresponding Author's Email: ravichemistry.kumar@gmail.com.

In: The Essential Guide to Alkaloids
Editor: Deepak Kumar Semwal
ISBN: 979-8-88697-456-0

1. Introduction

The increase in the world's population results in an increase in the need for agricultural and food production (Gerwick and Sparks, 2014). However, insect pests and diseases can negatively affect crop yields, which in turn affects the supply of food. In order to enhance crop yields and real output, a range of plant protection chemicals are created, produced, and used in agriculture. Chemicals used in plant protection are those that shield crops from biological threats such as diseases and insect pests. In the agriculture industry, plant protection agents are commonly referred to as pesticides, a product used to reduce the number of pests, any living organism that could harm or afflict crops (Koli et al., 2019). Though, because of their negative impacts on human health and the environment, these chemical pesticides are thought to be harmful in long run and viewed in a negative light by most of the developed and developing countries (Elfikrie et al., 2020, Sheikh et al., 2020, Isman and Grieneisen, 2014).

In view of the tremendous negative impacts of chemical pesticides, effective and environmentally friendly pest management methods are urgently desired to protect the environment and human health. Therefore, researchers continue in making efforts to find alternative natural-based products such as botanical pesticides to be used in plant protection. Botanical pesticides are obtained from different parts of plants of various plant families and essential oils. extracts and oleoresins extracted from the plant parts can also be used. The use of botanical pesticides is growing in popularity day-by-day due to their low toxicity, quick decomposition, and environmental friendliness (Wu et al., 2021).

One of the well-known and most diverse classes of natural products of pharmacological and toxicological significance is the alkaloid family, which has more than 20,000 identified and/or alkaloids. Insects that are vital for agriculture and public health have long been controlled using plant extracts that contain insecticidal alkaloids as bioactive components. Alkaloids are defined as "a cyclic organic compound containing nitrogen in a negative oxidation state which is of limited distribution among living organisms". Mostly, alkaloids are basic in nature and contain at least one tertiary nitrogen in their heterocyclic ring except for some exceptions such as paclitaxel, colchicine and caffeine. Alkaloids are mainly derived from biosynthetic amino acids for example phenylalanine, tryptophan, tyrosine, ornithine, and lysine. Mainly alkaloids are found and isolated from plants, however, some alkaloids have also been found in microorganisms, marine organisms like algae, dinoflagellates, puffer fish, and terrestrial animals i.e., insects, salamanders, and toads. Alkaloids are frequently categorized based on the structure of their molecules; for example, the two biggest classes are indole alkaloids and isoquinoline alkaloids. Other significant classes include steroidal alkaloids, pyridine and pyrrolizidine alkaloids, tropane alkaloids, etc. In some contexts, botanical origin-based classification is also used. For instance, cactus alkaloids, Catharanthus alkaloids, Cinchona alkaloids, Ergot alkaloids, Papaver (opium) alkaloids, Rauvolfia alkaloids, Solanum alkaloids, and Strychnos alkaloids, etc (Laboratories, 2005). Based on the facts about the huge diversity in chemical classes and pesticidal properties of alkaloids, this chapter is focusing on the role of alkaloids in plant protection as botanical pesticides.

2. Pesticidal Activities of Alkaloids

The pesticidal activities of alkaloids are mainly reflected in antifungal, antiviral, antibacterial, insecticidal, nematicidal and herbicidal activities, as discussed below (Figure 1).

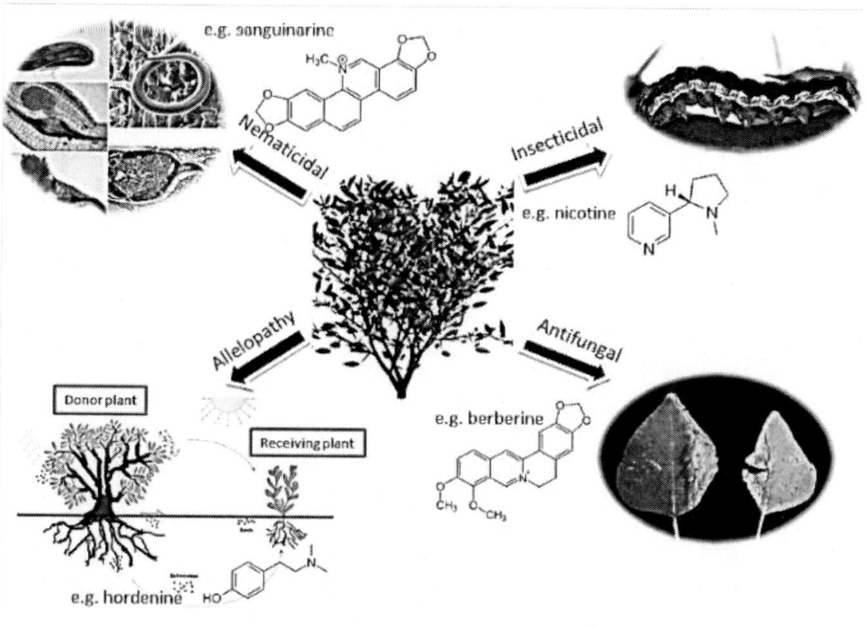

Figure 1. Pesticidal properties of alkaloids.

2.1. Insecticidal Activity

Plants employ alkaloids as toxic agents to combat herbivores, as they are toxic and cause death even in very low amounts. Alkaloids from different plants exhibit insecticidal properties against various insect species by causing a range of toxic effects such as malfunction of the midgut, redox imbalance, disruptions of biological membranes, inhibition of cholinesterase, disturbing development and metabolism, reproductive and acute toxicity (Chowanski et al., 2016). There are numerous alkaloids known for their insecticidal potential with a different mode of action. Nicotine, a natural insecticide present in the tobacco plant (*Nicotiana tabacum*) has been used as an insecticide since the 1690s, in the form of tobacco extracts and is the precursor for a large number of synthetic neonicotinoid insecticides. (Tomizawa and Casida, 2005) Nicotine and its derivatives mimic the action of acetylcholine and act as an

Table 1. Insecticidal activities of some alkaloid compounds

Plant species/ Family	Substance (Alkaloid)	Insect	Activity	EC_{50}/LC_{50} values	Ref.
Evodia rutaecarpa	Evodiamine [1], rutaecarpine [2]	*Aedes albopictus* (4th-instar larvae)	larvicidal	EC_{50}= 12.51 and 17.02, µg/ mL, respectively	Liu et al., 2012
Toddalia asiatica	Flindersine [3]	*Culex quinque-fasciatus, Anopheles stephensi*	larvicidal	EC_{50}= 568.88 ppm and 443.04 ppm	Duraipandiyan et al., 2015
Amaryllidaceae	Lycorine [4]	*Plutella xylostella*	larvicidal	LC_{50} =10.6 mg/L.	Liu et al., 2017
Solanaceae	Anabasine [5]	*Leptinotarsa decemlineata* Say	antifeedant	LC_{50} = 50 µg/cm^2	Singaravelan et al., 2015
Piper longum	Pipernonaline [6], piperoctadecalidine [7]	*Myzus persicae* S.	Insectici-dal	LC_{50} = 530 and 235 ppm, respectively.	Park et al., 2002
Tylophora indica	Tylophorinine [8]	*Helicoverpa armigera*	larvicidal	LC_{50} = 7.1 ppm	Kathuria et al., 2013
Penicillium simplicissimum AK-40	Okaramine A [9] and B [10]	Silkworm (3rd-instar larvae)	Insectici-dal	LC_{50} = 3 µg/g and 0.1 µg/g per diet respectively	Hayashi et al., 1989

agonist by readily interacting with the nicotinic acetylcholine receptors (nAChRs), resulting in the excitation and paralysis of insects (Taillebois et al., 2018). Alkaloids such as capsaicin, cassaine, nitidine and solenopsine are well-known inhibitors of Na^+/K^+-ATPase. Other alkaloids that bind to the neuro-transmitter receptors are nicotine, anabasine, morphine, strychnine, muscarine, pilocarpine, etc (Wink, 1998). Glycoalkaloids extracts (α-chaconine and α-solanine) affect heart contractile activity in insects and are found to be toxic to leaf-eating insects (Marciniak et al., 2010). The quinolizidine alkaloids kill insects by inhibiting protein synthesis in insects, affecting the potassium channel and glutamate receptor (Ujvary et al., 1991). The insecticidal properties of some alkaloid compounds are enlisted in Table 1 and the structures of compounds are illustrated in Figure 2.

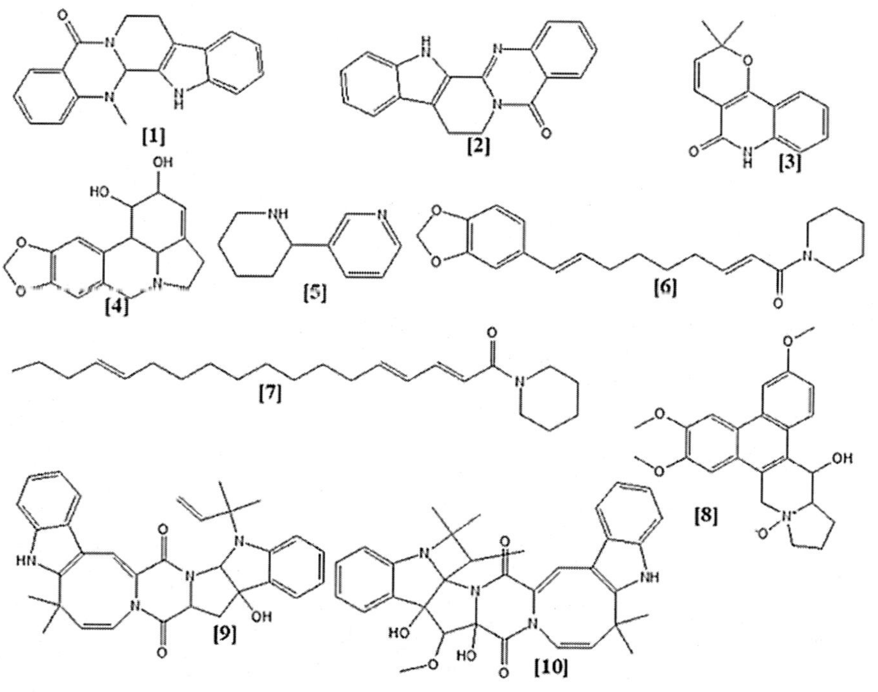

Figure 2. Structures of alkaloids having insecticidal activities.

2.2. Nematicidal Activity

Plant-parasitic nematodes are major pests affecting crop productivity and production. However, they do not cause direct crop loss, they come into play by forming disease complexes with other pathogens and micro-organisms (Singh et al., 2015). Botanical nematicides constitute a key component of a pest management system. Chemical classes of botanical nematicides include alkaloids, aldehyde and ketones, glycosides, glucosinolates, limonoids, saponins, phenolics, quinones and terpenoids (Ntalli and Caboni, 2012). Over 4000 plant parasitic nematode species have been identified, most of them feeding on the root, but some also on the aerial parts. Among the most notorious plant pathogenic nematodes, *Meloidogyne* sp., are found in host tissues and protected areas of soil. *Globodera* and *Heterodera* species (cyst nematode) are another important plant parasitic nematode. A number of studies have confirmed the effect of alkaloids against the root-knot nematode as well as on other nematode species as illustrated in Table 2 and the structures of the compounds illustrated in Figure 3.

Table 2. Alkaloids effective against different plant pathogenic nematode species

Alkaloid	**Effective against Nematode**	**LC_{50}/ EC_{50}/ IC_{50}/ LD_{50} values**	**Ref.**
Sanguinarine	*Bursaphelenchus xylophilus*	LC_{50} = 28.52 μg/mL	Kui et al., 2012
	Caenorhabditis elegans	LC_{50} = 22.78 μg/mL	
	Meloidogyne incognita	LC_{50} = 67.52 μg/mL	
Chelerythrine	*B. xylophilus*	LC_{50} = 34.50 μg/mL	Kui et al., 2012
	C. elegans	LC_{50} = 40.25 μg/mL	
	M. incognita	LC_{50} = 61.0 μg/mL	
Allocryptopine	*B. xylophilus*	LC_{50} = 37.45 μg/mL	Kui et al., 2012
	C. elegans	LC_{50} = 38.90 μg/mL	
	M. incognita	LC_{50} = 76.56 μg/mL	
Waltherione A	*M. incognita*	EC_{50} = 0.18 μL/mL	Jang et al., 2015
Waltherione E	*M. incognita*	EC_{50} = 0.08 μL/mL	Jang et al., 2015
Drupacine	*M. incognita*	-	Wen et al., 2013
	B. xylophilus		
Hyoscyamine	*M. incognita*	LC_{50} = 182.4 μg/mL	Babaali et al., 2021
Nicotine	*C. elegans*	LD_{50} = 46.1 μg/mL	Yu and Potter, 2008
	Heterodera glycines	LD_{50} = 4.5 μg/mL	
	H. schachtii	LD_{50} = 37.3 μg/mL	
	Meloidogyne hapla	LD_{50} = 3.3 μg/mL	
	M. incognita	LD_{50} = 51.6 μg/mL	
	Pratylenchus crenatus	LD_{50} = 3.2 μg/mL	

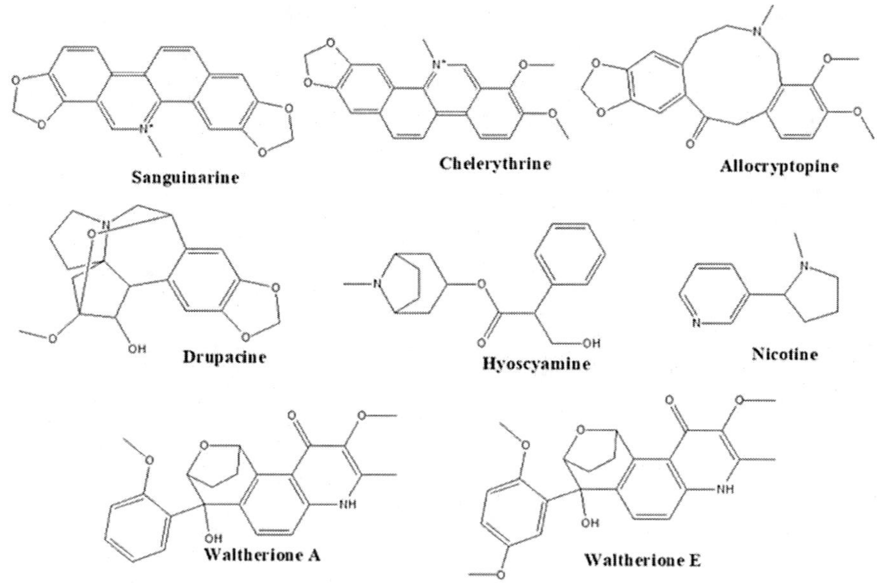

Figure 3. Structures of alkaloids having nematicidal activities.

2.3. Anti-Fungal Activity

Fungi are among the most important causal organism of plant diseases that causes significant crop losses. Fungi use a different mechanism to cause diseases; some kill their host tissues and feed on decay material, while some colonize the living tissues. Some fungus produces mycotoxins, which are extremely poisonous to plants. Those fungi include *Claviceps, Aspergillus, Fusarium, Trichoderma, Penicillium,* etc. Several alkaloid compounds are known to be effective against plant pathogenic fungi even at very low concentrations. The antifungal mechanisms of alkaloids are as follows (Lee et al., 2014, Thu et al., 2018).

Securinine and allosecurinine isolated from the methanol extract of *Phyllanthus amarus* exhibit the spore germination inhibition of several pathogenic fungi including *Alternaria alternata, A. brassicae, A. brassicicola, Erysiphe pisi, Colletotrichum musae, Colletotrichum sp., Curvularia lunata, C. pallenscens, C. maculans, Helminthosporium echinoclova* and *H. spiciferum* (Singh et al., 2008). Many isoquinoline alkaloids such as berberine, chelidonine, sanguinarine, and chelerythrine possess a wide range of antifungal activities. Sanguinarine enhanced the reactive oxygen species

(ROs) production, which might destroy the cell membranes by destroying the phospholipid residues and mitochondrial membrane by peroxiding (Zhao et al., 2019). Antifungal activities of some alkaloids against various plant pathogenic fungi are illustrated in Table 3 and their structures are in Figure 5.

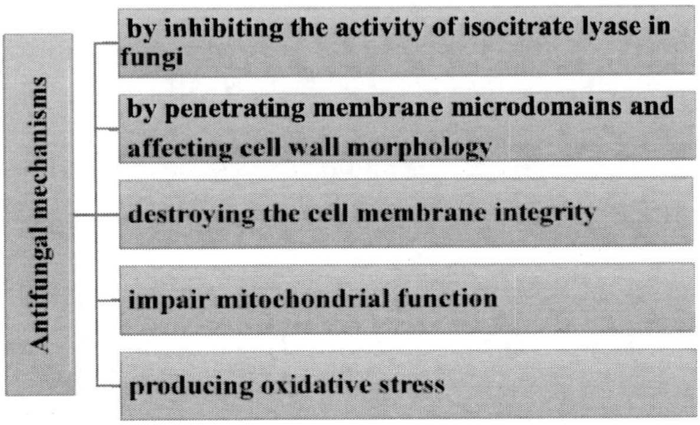

Figure 4. Different antifungal mechanisms shown by alkaloids.

Table 3. Antifungal activities of some alkaloid compounds

Alkaloid	Source	Fungal strains inhibited	Ref.
Venenatine	*Alstonia venenata*	*Fusarium udum, Ustilago cynodontis, Aspergillus flavus, Alternaria brassicicola*	Singh et al., 2000
d-calycanthine and l-folicanthine	*Chimonanthus praecox*	*Exserohilum turcicum, Bipolaris maydis, Sclerotinia sderotiorum, Alternaria solani, Fusarium oxysporum.*	Zhang et al., 2009
Stephanine and crebanine	*Stephania dielsiana*	*Cercospora kaki, Rhizoctonia solani, Pyricularia oryzae, Colletotrichum graminicola, Gymnosphorangium haraeanum*	Deng et al., 2011
Berberine iodide	*Fumaria indica*	*Curvularia lunata, Erysiphe cichoracearum, E. pisi, Fusarium udum, Penicillium sp.*	Sarma et al., 1999
Norsanguinarine	*Eschscholtzia californica*	*Alternaria brassicola, Fusarium lini, Helminthosporium pennisetti, Alternaria brassicae, Curvularia maculans*	Singh et al., 2009
Sanguinarine	-	*Magnaporthe oryzae*	Zhao et al., 2019
N1-decarbomethoxy chanofruticosinic acid	*Kopsia hainanensis*	*Rhizoctonia solani, Penicillium italicum, Fusarium oxysporum*	Chen et al., 2014

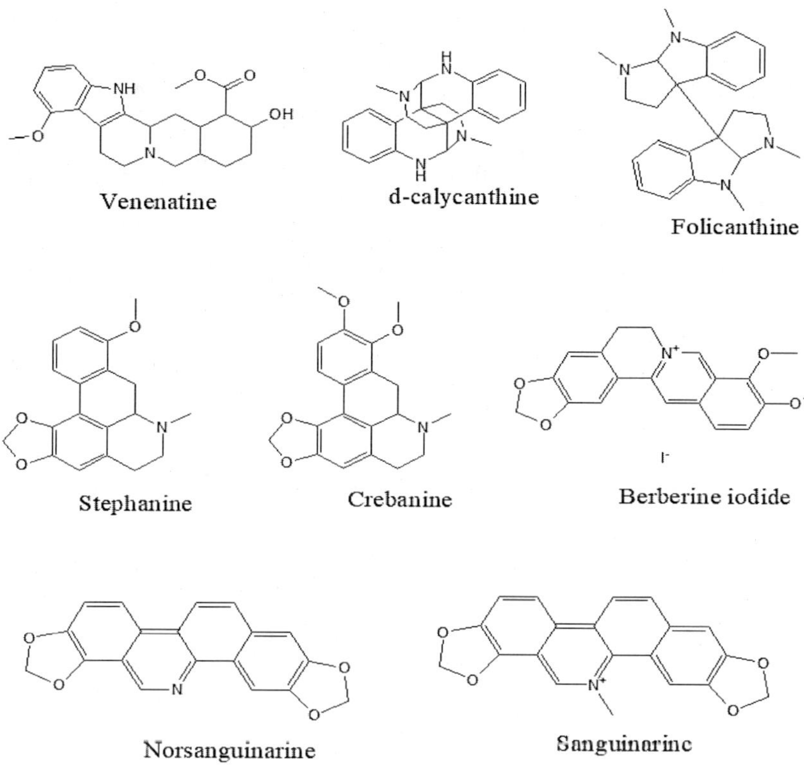

Figure 5. Structures of alkaloid compounds with antifungal activity.

2.4. Herbicidal and Phytotoxic Activities

Weeds directly compete with crops for nutrients, moisture, and light, which limits agricultural productivity (Colbach et al., 2017). Weeds also act as hosts for insects, nematodes, and fungal infections, which significantly slow down the development rates and biomasses of crops. Herbicidal sprays have helped sustainably to boost crop yields and may be the most effective and cost-effective weed management method available. However, using herbicides increases the risk to human health and pollutes the environment and the development of herbicide-resistant weeds is another impact of using herbicides. Some weed species including *Echinochloa crusgalli, Alopecurus aequalis* and *Amaranthus retroflexus* have become resistant to herbicides like glyphosate (Feng et al., 2019). Therefore, it is necessary to develop an eco-friendly, cost-effective, and effective weed management alternative to pesticides. Alkaloids and other natural compounds possess allelopathic properties, these compounds kill neighbouring plants and weeds when released into the environment by preventing the germination and growth processes (Badmus and Afolayan, 2012). Compounds isolated from the extracts of *Piper sarmentosum* (Piperine, sarmentine, sarmentosine, pellitorine) exhibit phytotoxic and herbicidal activity against the barnyard grass (*Echinochloa crusgalli*) and *Amaranthus retroflexus* (Feng et al., 2019). The alkaloid compound, holadysenterine isolated from a plant pathogenic fungus, *Drechslera australiensis* displayed promising herbicidal activity against *Rumex dentatus* L. weed (Akbar et al., 2014). Berberine is a natural herbicide compound isolated from *Coptis chinensis,* which showed herbicidal activity similar to the commercially available herbicide cinmethylin against *Bidens pilosa* (Wu et al., 2017). Some alkaloids like arborinine, graveoline isolated from *Ruta graveolens* L. demonstrate potent phytotoxic activity against *Spinacea oleracea* L. by inhibiting the electron transport chain process (Sampaio et al., 2018). The structures of compounds exhibiting herbicidal activity are shown in Figure 6.

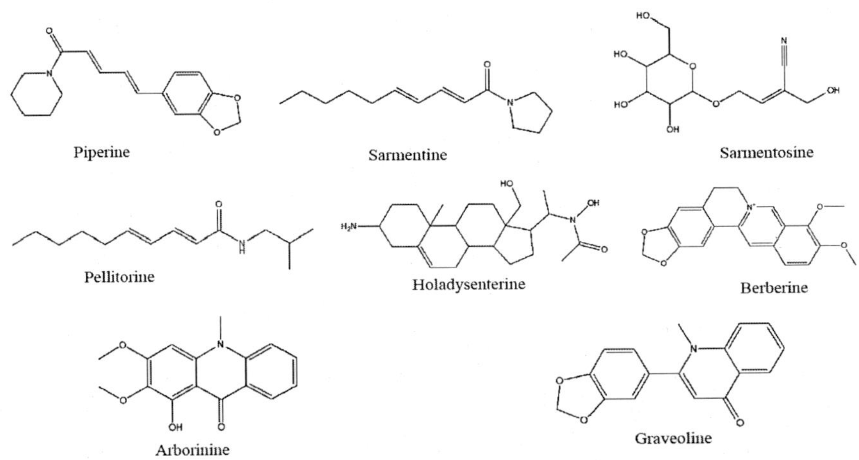

Figure 6. Structures of compounds with herbicidal activity.

2.5. Antibacterial Mechanism of Alkaloids

Bacteria are thought to have a simple structural design, but they have a wide range of metabolic capabilities. They are abundant almost everywhere on

Earth, from thriving in hydrothermal vents deep beneath the ocean's surface to living in jet fuel and on the rims of volcanoes. Animal digestion, fixation of nitrogen in the roots of legumes, the breakdown of animal and plant wastes and sewage disposal systems are just a few of the processes aided by beneficial bacteria. Both harmful and beneficial bacteria exist. On the other side, pathogenic bacteria result in severe, frequently fatal diseases in people, animals and plants. *Erwina*, *Pectobacterium*, *Pseudomonas*, *Clavibacter*, *Xanthomonas* and *Xylella* are some of the genera that include the majority of the plant pathogenic bacteria. They cause numerous symptoms like galls and overgrowth, wilts, leaf spots, blights, cankers, etc (Williams et al., 2017). Alkaloids' distinctive chemical structures have led to the gradual discovery of the processes underlying their antibacterial actions. Alkaloids prevent bacterial growth in a number of ways which include, preventing the production of nucleic acids and proteins, altering the permeability of bacterial cell membranes, damaging bacterial cell walls and cell membranes, impairing bacterial metabolism and inhibiting efflux pumps (Yan et al., 2021).

The expression of virulence genes is prevented by damage to DNA/RNA molecules or inhibiting DNA replication, which affects the features of bacteria and their growth and reproduction. The formation of diaphragms and the formation of a ring structure at the division site, that eventually regulates the process of bacterial cell division, are both facilitated by a protein FtsZ called filamentous temperature sensitive protein Z, as a result, FtsZ is being used as a test subject for the new antibacterial substances (Yan et al., 2021).

The phospholipid bilayers and proteins that make up the bacterial cell membrane form a flexible, semipermeable membrane. It aids in cell identification and electron transport and offers an interior environment that is rather stable for bacterial life activities. A selective advantage for bacterial survival in challenging environmental settings is the capacity to produce biofilms. When the bacterial cell membrane is severely damaged, the cell's protective barrier is broken, allowing a huge number of molecules to escape. Further, the cell wall's defense function is lost, and the cell membrane's transport and information transfer functions are obstructed (Yan et al., 2021).

Bacterial transmembrane protein complexes are drug efflux pumps. The pathogenic bacterial cell stays away from the membrane's antibacterial agent if the efflux pump system successfully excludes membrane-permeable antibacterial agents on time. This reduces the antibacterial agent's bactericidal effect and increases the pathogen's resistance through the membrane.

Numerous pieces of research have demonstrated that *E. coli* biofilms display higher levels of drug resistance compared to planktonic cells. This is related to the expression of efflux pump genes being higher in biofilms, which might result in an increase in the distribution density or activity of cell membrane efflux pumps. While this process can result in the biofilm-forming strains developing drug resistance, the reduction of efflux pump protein expression can increase the amount of time that pathogenic bacteria and antibacterial drugs have in contact with the membrane, potentially resulting in sterilization (Yan et al., 2021).

Asperthrin A and E, the prenylated indole diketopiperazine obtained from marine-derived endophytic fungus *Aspergillus sp* exhibit potential antibacterial activity against the agricultural pathogenic bacteria *Xanthomonas oryzae* pv. *Oryzae* and *X. oryzae* pv. *Oryzicola* (Yang et al., 2021). Alkaloids extracted from the leaves of *Prosopis juliflora* possess potent antibacterial activity against the *Xanthomonas sp.* Viz., *Xanthomonas axonopodis* pv. *malvacearum, Xanthomonas axonopodis* pv. *phaseoli and Xanthomonas compestris* pv. *vesicatoria.* The MIC was found to be 6, 5 and 4 μg/mL for the three species respectively (Raghavendra et al., 2009).

Three alkaloids isolated from *Peganum harmala* namely, pehanine, the quinazoline alkaloid and harmol and harmine, the β-carboline alkaloids were found to possess moderate inhibition zones against certain phytopathogenic bacteria. The total alkaloid extract showed average inhibition zones of 7.67-26.67mm against the test pathogens, *Ralstonia solanacearum*, *Ervinia amylovora*, *Pectobacterium carotovorum* and *Burkholderia gladioli. Ralstonia solanacearum* was found to be highly sensitive to the alkaloid extract and its unusual cell damage and genome clotting were observed using a Transmission Electron Microscope at the lowest dose of 4 μg/mL along with disordered cytoplasm and thicker cell walls (Shaheen and Issa, 2020).

The imidazole alkaloids, fusaritricines, isolated from the endophytic fungi *Fusarium tricinctum*, display significant antibacterial activity against the pathogen *Pseudomonas syringae* pv. *actinidae* which severely destroys the Kiwi plant (*Actinidia chinensis*) causing the canker disease (Ma et al., 2022). Chelerythrine and Sanguinarine extracted from *Macleaya cordata* display activity against the bacterial pathogens *A. tumefaciens, P. lachrymans, X. vesicatoria* with minimum inhibitory concentration, of 8.0 μg/mL, 32.0 μg/mL, 16.0 μg/mL for Chelerythrine and 8.0 μg/mL, 16.0 μg/mL, 16.0 μg/mL for Sanguinarine (Liu et al., 2009).

2.6. Antiviral Activities

Alkaloids' action mechanisms vary depending on their chemical structure and class. Numerous studies and applications have been made of both isolated chemicals and plants (or extracts of those plants) that contain alkaloids. It has been demonstrated that alkaloids interact with a variety of biological targets. The principal antiviral activities of many widely used phytochemical substances include antioxidant potential, scavenging properties, inhibition of DNA and RNA synthesis, and viral replication blockage, all of which may be caused by the synergistic effects of more than one mechanism. Research on the structural and conformational changes caused by the interaction of viruses with small molecules is of great interest since nucleic acids are the most ideal targets for limiting the particular viral replication stage. Numerous alkaloids disrupt the biological processes of the enzymes and genetic material involved in the viral cycle through a variety of methods including inhibition of DNA and RNA synthesis, inhibition of protein synthesis, etc (Abookleesh et al., 2022).

Even though the cell membrane is not the primary target, some investigations suggested that numerous natural alkaloids can effectively interfere with it by adhering to the viral glycoprotein. Alkaloids' amphiphilic nature probably explains why they have antiviral properties, especially as an entrance inhibitor that prevents viral attachment (Abookleesh et al., 2022).

The tobacco mosaic virus (TMV) is the most well-known and ancient plant virus. It is known to affect at least 125 different species of plant, including ornamental flowers, tobacco, peppers, tomatoes and cucumber. Additionally, TMV can serve as a model virus for the development of an antiviral drug against other viruses. Chelerythrine, extracted from *Chelidonium majus* exhibit potent antiviral activity against the most known plant virus TMV. It successfully reduced the damage caused by tobacco which ascribe to direct and indirect actions on TMV. Another alkaloid, sanguinarine also possesses a moderate inhibitive effect on TMV. An inhibition of TMV in tobacco is also caused by chelidonine (Guo et al., 2021). Different derivatives of topsentin also possess higher antiviral activity against the TMV and are thereby becoming a new lead substance for antiviral research (Ji et al., 2019).

Bioactivity research reveals that essramycin, the first isolated triazolopyrimidine natural product obtained from *Streptomyces sp.*, exhibits potential activity against the virus TMV. The core structure unit of essramycin is the same as that of ribavirin which is an effective antiviral agent used broadly for the prevention of TMV disease (Wang et al., 2019). The structures of compounds exhibiting antiviral activity are shown in Figure 7.

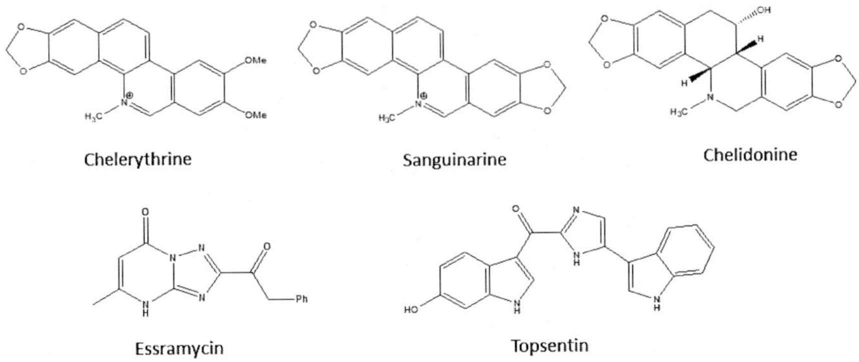

Figure 7. Structures of compounds having antiviral activity.

Conclusion

In recent years, the demand and development of botanical pesticides have increased in pest management strategies due to the negative side effects of conventional pesticides. A literature survey revealed that to a certain extent some alkaloid compounds are more effective than conventionally used pesticides. Moreover, any structural modification to the parent alkaloid enhances its biological potential. Furthermore, compared to synthetic pesticides, using alkaloids has benefits such as low toxicity, quick disintegration, and environmental friendliness, making these compounds desirable for further research in crop protection.

References

Abookleesh, F. L., Al-Anzi, B. S., & Ullah, A. (2022). Potential Antiviral Action of Alkaloids. *Molecules*, 27(3), 903.

Akbar, M., Javaid, A., Ahmed, E., Javed, T., & Clary, J. (2014). Holadysenterine, a natural herbicidal constituent from *Drechslera australiensis* for management of *Rumex dentatus*. *Journal of agricultural and food chemistry*, *62*(2), 368-372.

Babaali, D., Roeb, J., Zaidat, S., Reguige, B., Hammache, M., & Hallmann, J. (2021). Nematicidal activity of the tropane alkaloids hyoscyamine and scopolamine against the root-knot nematode *Meloidogyne incognita*. *Nematology*, *23*(7), 761-769.

Badmus, A., & Afolayan, A. (2012). Allelopathic potential of *Arctotis arctotoides* (Lf) O. Hoffm aqueous extracts on the germination and seedling growth of some vegetables. *African Journal of Biotechnology*, *11*(47), 10711-10716.

Chen, J., Yang, M. L., Zeng, J., & Gao, K. (2014). New broad-spectrum antibacterial and antifungal alkaloids from *Kopsia hainanensis*. *Phytochemistry Letters*, *7*, 156-160.

Chowański, S., Adamski, Z., Marciniak, P., Rosiński, G., Büyükgüzel, E., Büyükgüzel, K., & Bufo, S. A. (2016). A review of bioinsecticidal activity of Solanaceae alkaloids. *Toxins*, *8*(3), 60.

Colbach, N., Darmency, H., Fernier, A., Granger, S., Le Corre, V., & Messéan, A. (2017). Simulating changes in cropping practices in conventional and glyphosate-resistant maize. II. Weed impacts on crop production and biodiversity. *Environmental Science and Pollution Research*, *24*(14), 13121-13135.

Deng, Y., Yu, Y., Luo, H., Zhang, M., Qin, X., & Li, L. (2011). Antimicrobial activity of extract and two alkaloids from traditional Chinese medicinal plant *Stephania dielsiana*. *Food Chemistry*, *124*(4), 1556-1560.

Duraipandiyan, V., Baskar, K., Muthu, C., Ignacimuthu, S., & Al-Dhabi, N. A. (2015). Bioefficacy of flindersine against *Helicoverpa armigera* Hübner, *Spodoptera litura* Fabricius, *Anopheles stephensis* Liston. and *Culex quinquefasciatus* Say. *Brazilian Archives of Biology and Technology*, *58*, 595-604.

Elfikrie, N., Ho, Y. B., Zaidon, S. Z., Juahir, H., & Tan, E. S. S. (2020). Occurrence of pesticides in surface water, pesticides removal efficiency in drinking water treatment plant and potential health risk to consumers in Tengi River Basin, Malaysia. *Science of the Total Environment*, *712*, 136540.

Feng, G., Chen, M., Ye, H. C., Zhang, Z. K., Li, H., Chen, L. L., & Zhang, J. (2019). Herbicidal activities of compounds isolated from the medicinal plant *Piper sarmentosum*. *Industrial Crops and Products*, *132*, 41-47.

Gerwick, B. C., & Sparks, T. C. (2014). Natural products for pest control: an analysis of their role, value and future. *Pest Management Science*, *70*(8), 1169-1185.

Guo, W., Lu, X., Liu, B., Yan, H., & Feng, J. (2021). Anti-TMV activity and mode of action of three alkaloids isolated from *Chelidonium majus*. *Pest Management Science*, 77(1), 510-517.

Hayashi, H., Takiuchi, K., Murao, S., & Arai, M. (1989). Structure and insecticidal activity of new indole alkaloids, okaramines A and B, from *Penicillium simplicissimum* AK-40. *Agricultural and biological chemistry*, *53*(2), 461-469.

Isman, M. B., & Grieneisen, M. L. (2014). Botanical insecticide research: many publications, limited useful data. *Trends in plant science*, *19*(3), 140-145.

Jang, J. Y., Le Dang, Q., Choi, Y. H., Choi, G. J., Jang, K. S., Cha, B., & Kim, J. C. (2015). Nematicidal activities of 4-quinolone alkaloids isolated from the aerial part of *Triumfetta grandidens* against Meloidogyne incognita. *Journal of Agricultural and Food Chemistry*, *63*(1), 68-74.

Ji, X., Wang, Z., Dong, J., Liu, Y., Lu, A., & Wang, Q. (2016). Discovery of topsentin alkaloids and their derivatives as novel antiviral and anti-phytopathogenic fungus agents. *Journal of agricultural and food chemistry*, 64(48), 9143-9151.

Kathuria, V., Ruhl, S., Kaushik, N., Edrada-Ebel, R., & Proksch, P. (2013). Evaluation of bio efficacy of *Tylophora indica* leaf extracts, fractions and pure alkaloids against *Helicoverpa armigera* (Hübner). *Industrial Crops and Products*, *46*, 274-282.

Koli, P., Bhardwaj, N. R., & Mahawer, S. K. (2019). Agrochemicals: harmful and beneficial effects of climate changing scenarios. In *Climate change and agricultural ecosystems* (pp. 65-94). Woodhead Publishing.

Kui, W., Chao, L., Hao, L., Jianmei, X., Weibo, S., & Ligang, Z. (2012). Nematicidal activity of the alkaloids from *Macleaya cordata* against certain nematodes. *African Journal of Agricultural Research*, *7*(44), 5925-5929.

Laboratories, G. (2005). Definition and Classification of Alkaloids. Most, 56–61.

Lee, S. H., Moon, K., Kim, H., Shin, J., Oh, D. C., & Oh, K. B. (2014). Bahamaolide A from the marine-derived *Streptomyces* sp. CNQ343 inhibits isocitrate lyase in Candida albicans. *Bioorganic & Medicinal Chemistry Letters*, *24*(17), 4291-4293.

Liu, D. D., Guo, Y. F., Zhang, J. Q., Yang, Z. K., Li, X., Yang, B., & Yang, R. (2017). Inclusion of lycorine with natural cyclodextrins (α-, β-and γ-CD): Experimental and in vitro evaluation. *Journal of molecular structure*, *1130*, 669-676.

Liu, H., Wang, J., Zhao, J., Lu, S., Wang, J., Jiang, W., & Zhou, L. (2009). Isoquinoline alkaloids from *Macleaya cordata* active against plant microbial pathogens. *Natural Product Communications*, 4(11), 1934578X0900401120.

Liu, Z. L., Liu, Q. Z., Du, S. S., & Deng, Z. W. (2012). Mosquito larvicidal activity of alkaloids and limonoids derived from *Evodia rutaecarpa* unripe fruits against *Aedes albopictus* (Diptera: Culicidae). *Parasitology Research*, *111*(3), 991-996.

Ma, J. T., Du, J. X., Zhang, Y., Liu, J. K., Feng, T., & He, J. (2022). Natural imidazole alkaloids as antibacterial agents against *Pseudomonas syringae* pv. *actinidiae* isolated from kiwi endophytic fungus *Fusarium tricinctum*. *Fitoterapia*, 156, 105070.

Marciniak, P., Adamski, Z., Bednarz, P., Slocinska, M., Ziemnicki, K., Lelario, F. & Bufo, S. A. (2010). Cardioinhibitory properties of potato glycoalkaloids in beetles. *Bulletin of environmental contamination and toxicology*, *84*(2), 153-156.

Ntalli, N. G., & Caboni, P. (2012). Botanical nematicides: a review. *Journal of agricultural and food chemistry*, *60*(40), 9929-9940.

Park, B. S., Lee, S. E., Choi, W. S., Jeong, C. Y., Song, C., & Cho, K. Y. (2002). Insecticidal and acaricidal activity of pipernonaline and piperoctadecalidine derived from dried fruits of *Piper longum* L. *Crop protection*, *21*(3), 249-251.

Raghavendra, M. P., Satish, S., & Raveesha, K. A. (2009). Alkaloids isolated from leaves of *Prosopis juliflora* against *Xanthomonas* pathovars. *Archives of Phytopathology and Plant protection*, 42(11), 1033-1041.

Sampaio, O. M., Vieira, L. C. C., Bellete, B. S., King-Diaz, B., Lotina-Hennsen, B., Da Silva, M. F. D. G. F., & Veiga, T. A. M. (2018). Evaluation of alkaloids isolated from *Ruta graveolens* as photosynthesis inhibitors. *Molecules*, *23*(10), 2693.

Sarma, B. K., Pandey, V. B., Mishra, G. D., & Singh, U. P. (1999). Antifungal activity of berberine iodide, a constituent of *Fumaria indica*. *Folia microbiologica*, *44*(2), 164-166.

Shaheen, H. A., & Issa, M. Y. (2020). In vitro and in vivo activity of *Peganum harmala* L. alkaloids against phytopathogenic bacteria. *Scientia Horticulturae*, 264, 108940.

Sheikh, N., Patowary, H., & Laskar, R. A. (2020). Screening of cytotoxic and genotoxic potency of two pesticides (malathion and cypermethrin) on Allium cepa L. *Molecular & Cellular Toxicology*, *16*(3), 291-299.

Singaravelan, N., Nee'man, G., Inbar, M., & Izhaki, I. (2005). Feeding responses of free-flying honeybees to secondary compounds mimicking floral nectars. *Journal of chemical ecology*, *31*(12), 2791-2804.

Singh, A. K., Pandey, M. B., Singh, S., Singh, A. K., & Singh, U. P. (2008). Antifungal activity of securinine against some plant pathogenic fungi. *Mycobiology*, *36*(2), 99-101.

Singh, S., Jain, L., Pandey, M. B., Singh, U. P., & Pandey, V. B. (2009). Antifungal activity of the alkaloids from *Eschscholtzia californica*. *Folia Microbiologica*, *54*(3), 204-206.

Singh, S., Singh, B., & Singh, A. P. (2015). Nematodes: a threat to sustainability of agriculture. *Procedia Environmental Sciences*, *29*, 215-216.

Singh, U. P., Sarma, B. K., Mishra, P. K., & Ray, A. B. (2000). Antifungal activity of venenatine, an indole alkaloid isolated from *Alstonia venenata*. *Folia microbiologica*, *45*(2), 173-176.

Taillebois, E., Cartereau, A., Jones, A. K., & Thany, S. H. (2018). Neonicotinoid insecticides mode of action on insect nicotinic acetylcholine receptors using binding studies. *Pesticide biochemistry and physiology*, *151*, 59-66.

Thu, H. E., Hussain, Z., Mohamed, I. N., & Shuid, A. N. (2018). Recent advances in antibacterial, antiprotozoal and antifungal trends of *Eurycoma longifolia*: a review of therapeutic implications and future prospects. *Current Drug Targets*, *19*(14), 1657-1671.

Tomizawa, M., & Casida, J. E. (2005). Neonicotinoid insecticide toxicology: mechanisms of selective action. *Annual review of pharmacology and toxicology*, *45*, 247.

Ujvary, I., Eya, B. K., Grendell, R. L., Toia, R. F., & Casida, J. E. (1991). Insecticidal activity of various 3-acyl and other derivatives of veracevine relative to the Veratrum alkaloids veratridine and cevadine. *Journal of Agricultural and Food Chemistry*, *39*(10), 1875-1881.

Wang, T., Yang, S., Li, H., Lu, A., Wang, Z., Yao, Y., & Wang, Q. (2019). Discovery, structural optimization, and mode of action of essramycin alkaloid and its derivatives as anti-tobacco mosaic virus and anti-phytopathogenic fungus agents. *Journal of agricultural and food chemistry*, 68(2), 471-484.

Wen, Y., Meyer, S. L., Masler, E. P., Zhang, F., Liao, J., Wei, X., & Chitwood, D. J. (2013). Nematotoxicity of drupacine and a cephalotaxus alkaloid preparation against the plant-parasitic nematodes *Meloidogyne incognita* and *Bursaphelenchus xylophilus*. *Pest Management Science*, *69*(9), 1026-1033.

Williams, S. D., Boehm, M. J. & Hand, F. P. (2017). *Bacterial Diseases of Plants.* Department of Plant Pathology. Agriculture and Natural Resources. https://ohioline.osu.edu/factsheet/plpath-gen-6

Wink, M. (1998). Modes of action of alkaloids. In *Alkaloids* (pp. 301-326). Springer, Boston, MA.

Wu, J., Ma, J. J., Liu, B., Huang, L., Sang, X. Q., & Zhou, L. J. (2017). Herbicidal spectrum, absorption and transportation, and physiological effect on *Bidens pilosa* of the natural alkaloid berberine. *Journal of agricultural and food chemistry*, *65*(30), 6100-6113.

Wu, Y., Ren, D., Gao, C., Li, J., Du, B., Wang, Z., & Qian, S. (2021). Recent advances for alkaloids as botanical pesticides for use in organic agriculture. *International Journal of Pest Management*, 1-11.

Yan, Y., Li, X., Zhang, C., Lv, L., Gao, B., & Li, M. (2021). Research progress on antibacterial activities and mechanisms of natural alkaloids: a review. *Antibiotics*, 10(3), 318.

Yang, J., Gong, L., Guo, M., Jiang, T., Ding, Y., Wang, Z., & An, F., (2021). Bioactive indole diketopiperazine alkaloids from the marine endophytic fungus *Aspergillus* sp. YJ191021. *Marine Drugs*, 19(3), 157.

Yu, Q., & Potter, J. W. (2008). Selective nematicidal activity of nicotine. *Journal of Food, Agriculture and Environment*, *6*, 428-432.

Zhang, J. W., Gao, J. M., Xu, T., Zhang, X. C., Ma, Y. T., Jarussophon, S., & Konishi, Y. (2009). Antifungal activity of alkaloids from the seeds of *Chimonanthus praecox*. *Chemistry & biodiversity*, *6*(6), 838-845.

Zhao, Z. M., Shang, X. F., Lawoe, R. K., Liu, Y. Q., Zhou, R., Sun, Y., ... & Yang, C. J. (2019). Anti-phytopathogenic activity and the possible mechanisms of action of isoquinoline alkaloid sanguinarine. *Pesticide biochemistry and physiology*, *159*, 51-58.

Index

A

B

C

D

E

F

G

H

I

K

L

M

T

V

W

Z

About the Editor

Dr. Deepak Kumar Semwal is currently serving as Assistant Professor at Uttarakhand Ayurved University, India. He has been a Postdoctoral Fellow at Tshwane University of Technology (South Africa) and Panjab University (India). He received his MSc and PhD (Chemistry) from HNB Garhwal University (India). He published more than 120 research papers and 12 books.

Email: dr_dks.1983@yahoo.co.in